EPA420-B-06-010
July 2006

Questions and Answers on the Clean Diesel Fuel Rules

Assessment and Standards Division
Office of Transportation and Air Quality
U.S. Environmental Protection Agency

Introduction

The following are responses to questions received by the Environmental Protection Agency (EPA) concerning the manner in which the EPA intends to implement and assure compliance with the diesel fuel sulfur regulations at 40 CFR Part 80. This document was prepared by EPA's Office of Air and Radiation, Office of Transportation and Air Quality, and the Office of Enforcement and Compliance Assurance, Office of Regulatory Enforcement.

Regulated parties may use this document to aid in achieving compliance with the diesel fuel sulfur regulations. However, this document does not in any way alter the requirements of these regulations. While the answers provided in this document represent the Agency's interpretation and general plans for implementation of the regulations at this time, some of the responses may change as additional information becomes available or as the Agency further considers certain issues.

This guidance document does not establish or change legal rights or obligations. It does not establish binding rules or requirements and is not fully determinative of the issues addressed. Agency decisions in any particular case will be made applying the law and regulations on the basis of specific facts and actual action.

While we have attempted to include answers to all questions received to date, the necessity for policy decisions and/or resource constraints may have prevented the inclusion of certain questions. Questions not answered in this document may be answered in a subsequent document. The Agency intends to provide any additional responses as expeditiously as possible. Questions that merely require a justification of the regulations, or that have previously been answered or discussed in the preamble to the regulations have been omitted.

****New and Revised Questions****

The following questions have been revised and/or added since **September 2005**:

<u>Revised</u>
- 1.1, 1.4, 1.5, 1.6, 1.9, 1.42
- 2.26, 2.38, 2.45
- 3.16, 3.18
- 4.1, 4.6, 4.17, 4.27, 4.28, 4.34
- 7.1, 7.3, 7.6, 7.9, 7.11
- 9.2
- 11.7
- 13.1, 13.12, 13.16
- 15.4, 15.6, 15.7, 15.8
- questions revised to reflect the 45-day extension for downstream parties: 1.26, 2.4, 2.5, 2.13, 2.34, 2.38, 2.44, 12.18, 12.20, 15.6

<u>Added</u>
- 2.46, 2.47, 2.48, 2.49, 2.50
- 4.42, 4.43, 4.44
- 5.3
- 7.12
- 10.24
- 13.16, 13.17, 13.18, 13.19, 13.20, 13.21
- 14.8
- 15.10
- <u>Section 16</u>, which provides links to key regulatory documents (the Highway and Nonroad Diesel rules and related technical amendments, and a link to the electronic CFR)

The following questions have been revised and/or added since the **March 2005** version:

<u>Revised</u>
- *3.2, 3.3, 3.23, (also, the second 3.23 that appeared in the original QAs is now 3.24, 3.24 is now 3.25, and 3.25 is now 3.26)*
- *4.18, 4.35*
- *9.5*
- *10.7, 10.10, 10.14*

<u>Added</u>
- *1.29 - 1.50*
- *3.27*
- *4.36 - 4.41*

- *7.6 - 7.11*
- *9.9*
- *11.16, 11.17*
- *12.26, 12.27*
- *14.5 - 14.7*
- *15.9*

Table of Contents

List of Acronyms

ASTM	American Society for Testing and Materials
BOL	Bill of Lading
bpcd	Barrels per Calendar Day
CAA	Clean Air Act
CBI	Confidential Business Information
CDX	Central Data Exchange
CFR	Code of Federal Regulations
CI	Cetane Index
CTA	Credit Trading Area
D&T	Designate and Track
DTAB	Diesel (Fuel) Treated As Blendstock
EPA	Environmental Protection Agency
FTC	Federal Trade Commission
GPA	Geographic Phase-In Area
HS	High Sulfur (generally refers to sulfur levels greater than 500 ppm)
HW	Highway (also, *Motor Vehicle* or *MV*)
IRS	Internal Revenue Service
ISO	International Organization for Standardization
kbpd	Thousand Barrels Per Day
LDDV	Light-Duty Diesel Vehicle
LM, L&M	Locomotive and Marine Diesel
LSD	Low Sulfur Diesel (500 ppm sulfur)
mmBTU	Million British Thermal Units
MV	Motor Vehicle (also, *HW* or *highway*)
MVNRLM	Motor Vehicle, Nonroad, Locomotive, and Marine Fuel
MY	Model Year
NE/MA	Northeast/Mid Atlantic
NIST	National Institute of Standards and Technology
NOV	Notice of Violation
NR	Nonroad Diesel
NRLM	Nonroad, Locomotive, and Marine

OECA	Office of Enforcement and Compliance Assurance
OMB	Office of Management and Budget
OTAQ	Office of Transportation and Air Quality
PADD	Petroleum Administration Districts for Defense
PBMS	Performance Based Measurement System
ppm	Parts Per Million
PDD	Previously Designated Distillate
PTD	Product Transfer Document
SBA	Small Business Administration
SRM	Standard Reference Materials
RFG	Reformulated Gasoline
SR	Small Refiner
TCO	Temporary Compliance Option
ULSD	Ultra-Low Sulfur Diesel (15 ppm sulfur)
VAR	Volume Accounting Reconciliation
VCSB	Voluntary Consensus Standards Body

1. Designate and Track (D&T) Issues

1.1: When is EPA's Designate and Track (D&T) reporting mechanism expected to be completed?

> *A: D&T reporting will be an electronic reporting system, similar to the existing Tier 2 Gasoline Sulfur reporting program. Reports will be a flat file format based on spreadsheet templates. Additionally, we are working with EPA's Office of Environmental Information on a pilot project for accepting Confidential Business Information (CBI) through EPA's Central Data Exchange (CDX). CDX will enable us to utilize digital signatures and eliminate our dependency on a hybrid reporting system where electronic data must be accompanied by a hardcopy printout and certification with a "wet ink" signature. Completion of the CDX pilot is expected by August 2006.]*

1.2: We understand that many ULSD designated "facilities" are currently unregistered with EPA. They must apply for a registration or permit number by December 2005 in order to get a registration number by the NRLM rule's effective date. There is currently no application mechanism.

> *A: The RFG & Anti-dumping registration system will be the basis for ULSD registration and similar procedures will apply. Companies that have an existing Company ID with RFG & Anti-dumping may add ULSD facilities, while companies not registered with RFG & Anti-dumping will need to request Company and Facility IDs. Registration will be done using forms posted on OTAQ's Fuel and Fuel Additives Reporting Forms web page (http://www.epa.gov/otaq/regs/fuels/dieselfms.htm).*

1.3: Section 80.535 of the NRLM rule allows refiners to produce early credit NRLM low-sulfur (500 ppm) diesel. If undyed, this low-sulfur NRLM diesel is considered on-road fuel UNLESS it is designated as NRLM or LM diesel fuel. The NRLM early credit rule takes effect June of 2006. We will therefore need a registration number by June 1, 2006 in order to move early credit NR low-sulfur diesel fungibly with on-road. Is this correct?

> *A: Yes. Refinery designation requirements begin June 1, 2006, as well as several other provisions, including Product Transfer Documents (PTDs). Therefore, facility registration would be required prior to June 1, 2006. This is not limited to the NRLM early credit fuel.*

1.4: In addition, company software professionals will need at least 6 months to incorporate any external reporting system into the company software. What is the anticipated process and timing for developing the D&T reporting software? When will D&T software, or standards, be available to develop systems to feed them?

1

A: *EPA is not developing reporting software. The reporting process will be similar to the Tier 2 Gasoline Sulfur system – flat file, text or spreadsheet submission. Common practice is for EPA to share draft versions of forms with third party software vendors and industry representatives (AOPL, API, NPRA, etc.) for review and comment. Report Form instructions and templates for D&T are available on the OTAQ Fuel and Fuel Additives Reporting Forms web page (*http://www.epa.gov/otaq/regs/fuels/dieselfms.htm*).*

1.5: Most of the receipt and delivery facilities that pipelines operate do not have EPA registration numbers required to implement D&T. If a terminal receives product from several pipelines, each one will require a registration number for that custody transfer. Is EPA ready to process the volume of applications required? When will the registration applications be available?

A: *EPA will process ULSD facility registration applications in a timely manner and anticipates no difficulty in doing so. Registration will be done using the forms posted on OTAQ's Fuel and Fuel Additives Reporting Forms web page (which can be found at:* http://www.epa.gov/otaq/regs/fuels/rfgforms.htm*).*

1.6: If a pipeline does not handle 500 ppm highway fuel, it appears that they are still required to file D&T reports, is this correct? Systems cannot opt out of D&T by handling only certain products?

A: *All facilities are required to compile and file D&T reports if they handle any designated fuel upstream of the point where taxes are assessed, and dye and/or marker is added if required. This includes a pipeline that does not handle any 500 ppm highway diesel fuel. In that case, the D&T reports would confirm that the pipeline did not handle this highway fuel, and did not violate any of the restrictions on redesignating off-road fuel as highway fuel. The D&T reports from this pipeline would also be used in evaluating compliance by the parties before and after this pipeline in the distribution system. This pipeline is only one step in a chain of fuel hand-offs, and the D&T system is designed to use information from each party to a hand-off to ensure compliance by all of the parties in the chain of distribution.*

1.7: Will refiners be required to provide data sheets, or other types of testing data to insure pipelines receive the correct product?

A: *Refiners are required to maintain batch testing records which demonstrate compliance with the applicable sulfur standard. The PTD that accompanies the transfer of a batch of designated fuel from a refiner to a pipeline operator must correctly identify the sulfur standard to which the fuel complies. However, there is no EPA requirement that testing records accompany the batch of fuel when it is transferred to the custody of a pipeline operator. Any private requirements established between refiners and pipelines, would be business decisions that are separate from EPA's regulations.*

1.8:   What is the plain English clarification of § 80.598(b)(9)(vii)(B) and § 80.599(b)(5)?

A: *The regulations limit the volume of 500 ppm NRLM that may be redesignated as highway 500 ppm by fuel distributors. Section 80.598(b)(9)(vii)(B) applies to distributors other than truck loading terminals. To ensure that there are no significant redesignations over the annual compliance period for these facilities, the volume of 500 ppm sulfur highway diesel fuel that a facility discharges from it's custody must be no greater than 102 percent of the volume of 500 ppm sulfur highway fuel that the facility received during the annual compliance period, taking into account inventory change. This provides pipelines and other parties upstream of truck loading terminals a tolerance of 2% during the annual compliance period to account for metering differences, volume swell, and the like. See 69 Fed. Reg. 39070 (June 29, 2004). Section 80.599(b)(5) requires that diesel fuel greater than 500 ppm sulfur must be designated as a fuel other than 15 ppm or 500 ppm highway or NRLM. For example, it can be designated as heating oil or another diesel fuel that can lawfully be above 500 ppm. Requiring a proper designation for high sulfur diesel fuel is the first step in ensuring that it stays segregated from fuel that is supposed to meet the 15 ppm or 500 ppm sulfur level.*

1.9:   What are the Designate & Track Quarterly and Annual start and finish dates? In the Final Nonroad Diesel rule the Quarterly Compliance Periods listed in § 80.599 begin June 1, 2007 and end September 30, 2010, but the Quarterly Reporting listed in § 80.601 begins November 30, 2007 and ends August 31, 2010. Why would the report period end before the compliance period? There is a similar issue with the Annual report versus compliance dates.

A: *The dates for the quarterly and annual compliance periods and reporting dates were incorrect in the Final Nonroad Rule regulations but were correct in the preamble (69 FR 39100, June 29, 2004). This error was corrected in a technical amendment to the regulations (which can be found on EPA's web site at: http://www.epa.gov/otaq/regs/fuels/diesel/diesel.htm#regs, July 7, 2005).*

Following this, it was found that there was a 'gap' in when the red dye will not be required at the refinery gate and when the D&T compliance periods (and corresponding report dates) began. As a result, we published subsequent technical amendments to the regulations (the amendments were published in the Federal Register on November 22, 2005 and can be found at: http://a257.g.akamaitech.net/7/257/2422/01jan20051800/edocket.access.gpo.gov/2005/pdf/05-22807.pdf).

The dates for the compliance periods and reporting dates are as follows:

Quarterly Compliance Periods and Reporting Dates	
Quarterly Compliance Period:	*Report Due Date:*
June 1, 2006 – September 30, 2006	February 28, 2007
October 1, 2006 – December 31, 2006	
January 1, 2007 – March 31, 2007	August 31, 2007
April 1, 2007 – May 31, 2007	
June 1, 2007 – September 30, 2007	November 30, 2007
October 1, 2007 – December 31, 2007	February 28, 2008
January 1, 2008 – March 31, 2008	May 31, 2008
April 1, 2008 – June 30, 2008	August 31, 2008
July 1, 2008 – September 30, 2008	November 30, 2008
October 1, 2008 – December 31, 2008	February 28, 2009
January 1, 2009 – March 31, 2009	May 31, 2009
April 1, 2009 – June 30, 2009	August 31, 2009
July 1, 2009 – September 30, 2009	November 30, 2009
October 1, 2009 – December 31, 2009	February 28, 2010
January 1, 2010 – March 31, 2010	May 31, 2010
April 1, 2010 – May 31, 2010	August 31, 2010
June 1, 2010 – September 30, 2010	November 30, 2010

Annual Compliance Periods and Reporting Dates	
Annual Compliance Period:	**Report Due Date:**
June 1, 2006 - May 31, 2007	August 31
June 1, 2007 - June 30, 2008	August 31
July 1, 2008 - June 30, 2009	August 31
July 1, 2009 - May 31, 2010	August 31
June 1, 2010 - June 30, 2011	August 31
July 1, 2011 - May 31, 2012	August 31
June 1, 2012 - June 30, 2013	August 31
July 1, 2013 - May 31, 2014	August 31
July 1, 2014 - June 30, 2015	August 31

1.10: Is downstream D&T no longer required in 2010 or in 2015 or ever?

> *A:* *The downstream D&T provisions change over time, as the standards change and more of the highway and off-highway diesel fuel becomes subject to the 15 ppm refinery gate standard. In general, downstream parties will always have to designate the diesel fuel they transfer, as part of the PTD requirements. However over time there will be less need to track various kinds of diesel fuels and fewer volume balances that will apply. After May 2014 the only tracking and related volume balances that will apply downstream involve heating oil and 500 ppm LM, to avoid improper transfer of low sulfur heating oil into the LM market.*

1.11: If only credit fuel is sold into the heating oil market, does a marker need to be added? Does a pipeline need to track if not receiving heating oil into its system? Same for the marker in L&M, if only credit 500 ppm NR is received and no 500 ppm L&M, does it need to mark and does it need to track?

> *A:* *First, the requirement to mark heating oil is limited to heating oil used in certain parts of the country. In general, the marker does not need to be added to diesel fuel distributed from a terminal within the northeast and mid-Atlantic area and Alaska, for use in those areas. That means the great bulk of the heating oil market is not subject to a marking requirement. For heating oil produced or distributed outside those areas, the terminal is required to mark heating oil prior to distribution from the truck loading terminal. If the fuel is designated as something other than heating oil, then it does not need to be marked, and will be subject to the sulfur level applicable to its designation. Outside the areas noted above, any diesel fuel that is not marked is presumed to not be heating oil, and*

typically will be considered motor vehicle or NRLM fuel and subject to the appropriate sulfur standard.

Therefore, if it is designated as NRLM, it is not treated as heating fuel, but rather NRLM and it does not need to have the marker added. NRLM can always be used as heating oil or LM diesel fuel, without any change in its designation , however it remains subject to the sulfur standard that applies to its NRLM or LM designation. If the NRLM is redesignated as heating oil, it would need to be marked at the terminal. Similarly, if it is designated as NR fuel, and not L&M, then no marker needs to be added unless it is redesignated as LM.

Any pipeline that receives designated fuels (including high sulfur NRLM, and credit 500 ppm NR fuel) must comply with the designate and track requirements. This information is essential for downstream parties to be able to demonstrate their compliance with the marker requirements for heating oil and 500 ppm LM fuel (from 2010-2012). It will also be used by EPA to verify the compliance of parties who receive fuel from the subject pipeline by allowing us to compare the volumes of designated fuels reported as received from the pipeline with the volumes reported delivered by the pipeline.

1.12: Do pipelines without truck loading terminals need to submit quarterly D&T reports or just annual D&T reports?

A: *Pipelines without truck terminals must submit both annual and quarterly reports prior to July 1, 2010. The quarterly reports will contain the information on receipts and deliveries of designated fuel volumes. The annual reports will contain information to show compliance with volume balance requirements over the annual compliance period. Beginning, July 1, 2011, only annual reports need to be submitted (see chart on quarterly and annual compliance dates, in question 1.9), and compliance with volume balance requirements continues to be on an annual basis.*

1.13: In the definition of facility, can a common carrier pipeline be aggregated with any other non-common carrier terminals if they are all owned by the same parent company?

A: *Yes, if certain conditions are met. First, a single entity, in this case, the parent company must maintain custody of the fuel at all times in the aggregated facility (i.e. the parent company must maintain custody at all subsidiary/leased terminals defined as part of that aggregated facility). If the parent company/entity fails to maintain custody of the fuel at any point, that point cannot be considered part of the aggregated facility. In addition, the choice to treat places as aggregated or separate facilities may not be changed by the entity during any applicable compliance period. If a refinery is part of the aggregated facility, the common carrier pipeline may __not__ be included as part of the aggregated facility.*

6

1.14: If an *entity*, for a *facility* (as those terms are defined in 40 C.F.R. § 80.502), chooses a single registration number for its refinery, pipeline, and terminal (thus creating an 'aggregated facility'), does this mean that in the event of finding product in violation of the diesel fuel regulations, only the registered entity will be subject to civil penalty, and not all 3 portions of the system separately?

> A: *It is dependent on the type of violation. For volume balances, liability falls on the entity in relation to its facility. The volume balance is calculated over the entire aggregated facility that is registered, and the entity that registered the facility will be subject to the penalty. For other violations (such as violations of the applicable sulfur level of a batch of fuel), the liability is not limited. The fact that the terminal is aggregated with a pipeline and refinery for volume balance purposes is only relevant to this violation because it means, by the definition of an aggregated facility, that the entity has custody of the fuel at all locations. However any other entity or person who takes custody of the fuel may also be subject to presumptive liability.*

1.15: At the terminal, if a pour point depressant, conductivity improver, lubricity improver which is compliant to the 15 ppm standard is injected, must the terminal operator register as a refiner?

> A: *Injecting such additives to a batch of fuel would not make a terminal operator a refiner.*

1.16: Some pipeline and marine terminals receive fuel from multiple refineries. Will they need separate facility ID numbers?

> A: *The terminal owner must obtain a facility ID for its facility, subject to any ability to lawfully aggregate into a larger facility. The fact that the terminal facility receives fuel from various refineries does not change the fact that it is a still a single place in the distribution system that is owned by a single entity. Similarly, a pipeline receives a single ID for a single facility notwithstanding how many refineries supply fuel to that facility. Consequently, the terminal in this situation would need only one facility ID number for its terminal operations. However, if fuel is imported into the terminal, or the terminal acts as a refiner, the terminal will need to state such activity, in addition to the other quarterly reporting requirements of § 80.601, and additionally will have annual reporting requirements of § 80.604. The party will also need to register as a refiner or importer, and meet all applicable refiner or importer requirements, but that would not change the facility ID that goes with the terminal facility.*

1.17: If a company has two (or more) refineries and transfers highway diesel blendstocks between refineries, are they precluded from aggregating?

A: Refineries are precluded from aggregating with other refineries.

1.18:   If a terminal has 15 ppm No. 2 and 500 ppm kerosene (No. 1), can a truck blend the two (winter blending) and take 500 ppm to the pump at retail? Who is responsible for the downgrading accounting?

A: The two fuels can be blended as long as the appropriate volume balances are met. When the terminal received the 15 ppm No. 2, it would have been designated as highway or NRLM. The limit on downgrading applies to No. 2 15 ppm highway diesel fuel that the terminal receives, and restricts the ability of the terminal to redesignate it as 500 ppm highway. Both the terminal and the truck transporter would have to meet the anti-downgrading and any other applicable volume balances, based on comparing the volumes and designations of fuel they received to fuel delivered. Since accounting responsibility is keyed to what you received and delivered, it will make a difference whether the blending occurred before or after custody was transferred. For example, if the blending occurs before the terminal transfers custody, then the terminal has to account for the blended fuel in its delivery accounting. If blending occurred after the transfer of custody, then the truck transporter will have to account for the blended fuel in its delivery accounting.

1.19:   In attempting to balance reporting requirements, if a pipeline ticket is generated for a particular amount of barrels/gallons, a terminal will have a natural gain or loss. Do you have a tolerance or must it match exactly?

A: The reported hand-off volumes must match – there is no tolerance. However, to account for gains and losses in the system, the D&T provisions allow for a 2% gain/loss over the course of the annual compliance period for pipelines. See answer to question 1.8, above. For terminals, there is no such specific tolerance, however, they have the ability to adjust by redesignating volumes accordingly.

1.20:   Who is responsible for the product if it comes in to the terminal (from a pipeline) over 15 ppm?

A: Presumptive liability under 40 C.F.R. § 80.612 applies to the party in possession of the contaminated fuel (in this case, the terminal after it has taken custody) and to all parties upstream of the facility where the violation is found (including, in this case, the pipeline). Each party has an opportunity to establish a defense to liability. Until it is proven which party is at fault, all parties involved in the production and distribution of the product will be presumed liable.

1.21:   Please confirm that a terminal operator selling 500 ppm nonroad diesel outside the (Northeast/Mid-Atlantic) NE/MA Area does not have to add the marker because that fuel need not be differentiated from higher sulfur (greater than 500 ppm) nonroad diesel fuel generated by credits.

A: *Only diesel fuel designated as heating oil need be marked if it is distributed from either a terminal outside of the NE/MA Area or a terminal inside the NE/MA Area for delivery outside of the Area.*

Beginning in 2007, all heating oil must contain the marker outside the NE/MA area. From 2010 to 2012, fuel designated as 500 ppm LM fuel also must contain the marker. In this instance the terminal selling the 500 ppm fuel does not have to add a marker unless it is either heating oil or, during 2010 to 2012, LM fuel. If it is not marked, then it will be treated as subject to the appropriate sulfur standard. The terminal selling the high sulfur fuel would need to mark the fuel if it is heating oil. The high sulfur fuel could not be designated as 500 ppm LM, so the LM marking requirement should not apply.

1.22: A company owns a refinery and ships its product via a common carrier pipeline it owns (and now exclusively uses) to a terminal that it owns. Can these facilities be aggregated, or does the fact that the company's pipeline is a common carrier pipeline preclude aggregation?

A: *As stated in the regulation (§ 80.502(b)(1)), facilities may be aggregated if custody is maintained by one entity that owns those specific facilities. Per § 80.502(b)(2), refineries may only be aggregated with facilities that do not receive fuel from other refineries or import facilities. In the case stated here, the "common carrier" status of the pipeline does not matter if the pipeline is not receiving fuel from other entities. The common carrier pipeline example that was used in the regulations was only meant to refer to the typical case where an entity receives fuel from various sources, and thus would not meet the criteria for facility aggregation.*

1.23: If a company redesignates, please explain how to *restore* volumes so that required volume balances are met.

A: *The volume balance requirements limit certain types of redesignations, unless the volume balance is met over the compliance period. For example, during a compliance period a terminal may redesignate NRLM as highway fuel as long as over the compliance period the terminal meets the required volume balance. If a terminal has been receiving fuel designated as NRLM and delivering some of it as highway and some as NRLM, the volume balance on highway fuel could go negative during that compliance period. To offset this, the terminal would have to take delivery of the fuel designated as highway and redesignate enough of it as NRLM to satisfy the volume balance.*

9

1.24:   How will a terminal designate a mixture of transmix and "off-spec" fuel that might be a combination of 15 ppm, all grades of 500 ppm, heating oil, and jet fuel? Will this designation(s) limit the marketing options of a transmix handler <u>beyond</u> the limits that would be imposed strictly by the sulfur content of the mixture alone?

     A: *The designation is left to the terminal. The designation must be clear and accurate, therefore the sulfur level of the mixture will limit the designation options. Beyond that, the designate and track compliance calculations have been designed to minimize, if not eliminate, any unintended marketing limitations.*

1.25:   According to the preamble (page 39067), "*EPA intends to work with industry subsequent to this final rule to provide guidance regarding facility boundary and aggregation decisions that will address the many unique situations.*" It appears from the discussion in the preamble that such "unique situations" are present when the same entity owns and operates a series of locations in the distribution system. Under the rule, the entity may choose whether to aggregate or to treat the separate locations as separate facilities. It would appear that such guidance might not be necessary, because an entity has the flexibility under the rule to determine where to draw lines around its facilities, so long as each facility meets the definition of "facility" under the rule. What is EPA's intent with regard to such guidance? When can industry expect to become part of these discussions? Has EPA developed a working list of the "unique situations"? Is EPA's intent with the guidance to address the definition of a "facility" and to help the regulated community understand the parameters of a facility?

     A: *In addition to this Question/Answer document, EPA has released Small Entity Compliance Guides (see http://www.epa.gov/cleandiesel/comphelp.htm) to assist small entities in complying with the requirements of the rulemaking. It was our intention that the presentations and discussions that occurred at the Clean Diesel Fuel Workshop in November 2004 would help to clarify and address questions and/or concerns that industry had regarding the definition of "facility." We will also answer any other questions as they arise and will continue to work with the fuel industry to ensure a smooth transition to ultra low sulfur diesel fuels.*

1.26:   When will the new technology vehicles start being sold (i.e. what is the absolute target date for ULSD at the pump)?

     A: *Heavy-duty vehicle model years typically start on January 1. However, they could start several months earlier and medium-duty vehicles often do. Light-duty vehicle model years typically start even earlier. We designed the program with a retail target date of September 1, 2006 (and a recent technical amendment to the regulations (published November 22, 2005) extended this deadline to October 15, 2006). The regulations set out the dates when fuel must meet the sulfur standards at various points in the distribution system. The regulations also prohibit selling anything other than 15 ppm sulfur diesel fuel to 2007 or later model year highway vehicles.*

1.27: My company is co-owner of a single physical pipeline that is operated as if it were two separate pipelines. Each of the co-owners has a set fraction of the pipeline's capacity and has established its own tariff rates which it charges to its customers. May each of the co-owners obtain a separate facility registration under the designate and track system with respect to its operations on this pipeline?

    A: *Section 80.502(b)(5) of the diesel fuel regulations states that a contiguous pipeline may not be subdivided into more than one facility. Based on this, your pipeline could not be subdivided into two facilities with separate facility registrations. EPA adopted this requirement because we believe that allowing an individual pipeline to be subdivided for D&T facility registration purposes would severely impact EPA's ability to evaluate compliance and assign liability for potential violations.*

1.28: My company imports diesel fuel into storage tanks that we lease at a terminal in New England. Please explain the responsibilities of the importer and the terminal owner with respect to the designate and track requirements related to the importation of diesel fuel.

    A: *The importer must register separately for each terminal facility into which it imports fuels. EPA will issue an import facility registration number for the importer that is composed of a numerical entity code and a numerical location code (the location being the terminal). The terminal owner will be provided with a terminal facility registration number composed of a location number (same as that supplied to the importer) and an entity registration number. Alternately, if the imported fuel is certified before delivery to a U.S. terminal, then the importer may obtain a registration for a "virtual" off-shore import facility. In such cases, the importer will be given an import facility registration number applicable to the importation of fuel into any terminal located in a single PADD. It has been the practice for EPA to issue these types of "virtual" import facility registrations for imported gasoline.*

The importer must appropriately designate all imported fuels. For each of its registered facilities, the importer must also fulfill all of the record-keeping and reporting obligations applicable to an importer. For example, the importer must demonstrate compliance with the requirement that 80% of all highway diesel fuel it imports meets a 15 ppm sulfur standard (2006-2010).

The terminal owner must account for the volumes of designated fuels that are imported into its terminal in demonstrating compliance with the volume balance and anti-downgrading requirements.

11

1.29:   The regulations state that a contiguous pipeline may not be broken down into more than one facility for the purposes of complying with the designate and track reporting requirements. There is a break-out tank facility in our pipeline system where all of the fuel in the trunk line feeding the facility (trunk line #1) typically enters break-out tankage before being passed further down our system through another segment of our trunk line (trunk line #2) or to several stub-lines. However, it is possible for some or all of the fuel in trunk line #1 to bypass the subject break-out tank facility and to be fed directly into trunk line #2. Could trunk lines #1 and #2 be considered separate facilities?

   A: *The facilities described in the question could be registered separately. The provisions in § 80.502(b)(5) are only intended to prevent the subdivision of a pipeline for reasons not sufficiently related to a true physical division in a section of a pipeline owned by one entity, such as the presence of a booster pump station in the line. Break-out tank facilities are necessary at locations in a pipeline system where pipelines of different diameters and flow rates are connected, and where branching in the pipeline system takes place. Therefore, a break-out facility is an appropriate location at which to draw a distinction between different parts of the pipeline system with respect to compliance with the designate and track requirements.*

1.30:   My company owns certain distinct physical parts of a terminal (pipeline manifold, several storage tanks, and the truck loading rack), whereas other parts of the same terminal (several storage tanks) are owned by another company. Would EPA consider allowing such a terminal to be subdivided for registration purposes under the designate and track (D&T) provisions of the diesel sulfur regulations? Under such an approach, each party would register, keep records, and demonstrate compliance for those physical assets at the terminal which it owns. It would be inappropriate to force these parties to become jointly responsible for the terminal as a whole.

   A: *A facility is defined as the place or series of places at which an entity produces, imports, or maintains custody of the fuel, extending from the point of initial custody to the point that custody is transferred (§ 80.502(b)). Under this definition, there would be two facilities at the terminal described in the question- one associated with each owner/entity. Each facility would cover the part of the terminal where the owner/entity had custody of the fuel. This is also consistent with § 80.502(b)(5), which provides that an individual terminal can not be subdivided into more than one facility (without approval by EPA). That provision applies to the process of voluntarily subdividing a facility after the physical extent of the facility has been initially determined under § 80.502(b). This terminal subdivision provision does not determine the physical extent of the location that initially meets the definition of facility under § 80.502(b). It restricts the voluntary subdivision of the facility once the extent of the facility has been determined under § 80.502(b). In this case, the definition of facility would lead to the existence of two separate facilities. Section 80.502(b)(5) then means that neither of these two facilities may be subdivided without approval by EPA.*

It is important to note that this does not mean that joint ownership or an owner/tenant relationship in a terminal would result in the terminal having more than one facility. In such cases, the definition of facility leads to the initial determination that the entire individual terminal is a single facility for the owner/entity. In that case, approval of the Administrator would be required to voluntarily subdivide the terminal into more than one facility, under § 80.502(b)(5).

1.31: Will EPA please explain the registration and other requirements for marine vessel owners?

    A: EPA will release a separate QA/guidance information on D&T for marine vessels and other mobile operations.

1.32: Can you please describe the requirements for a diesel marketer picking up fuel at a terminal and making a delivery to a retail or wholesale purchaser-consumer site?

    A: If the fuel has already been dyed, taxed, and/or marked, there are no D&T requirements for the marketer. The diesel marketer does not need to register with EPA and meet the D&T requirements unless they intend to distribute MVNRLM 500 ppm diesel fuel on which taxes have not been assessed, NRLM fuel that is undyed, or heating oil or LM fuel that is not marked (in those areas of the country where the marker is required, per § 80.597(c)(1)). In either case, the marketer has PTD requirements and would still need to maintain the integrity of the fuel and, for defense purposes, should have an adequate quality control and sampling and testing programs. The marketer is also still subject to the anti-downgrading provisions of § 80.527.

1.33: Is it possible to receive fuel from/deliver fuel to someone who is not registered? The regulations state that everyone has to register, however what would happen in the case of an importer who has never imported fuel to the U.S., and does not plan to do so, but just so happens to import a small batch of fuel. Such a party will not likely be registered with EPA prior to importing the fuel.

    A: Per the regulations of § 80.597(c), all parties in the diesel fuel production and distribution system, up to the point where the fuel leaves the D&T boundaries (i.e., dyed or taxed) are required to register. However, it is possible to receive fuel from an unregistered entity or facility, as per the requirements and limitations of § 80.601(a)(3). For this case only, the entity/facility ID 8888-88888 must be used.

There is no provision to transfer fuel to an unregistered entity or facility in the D&T system (i.e., up to and including the truck loading terminal). In this case, the unregistered entity/facility must contact EPA to request ID numbers. In the case of the hypothetical importer in the question, it may transfer fuel into the D&T system without immediately registering (subject to the previously mentioned requirements and limitations of § 80.601(a)(3) on the recipient), but the importer must also comply with recordkeeping and reporting requirements, such as annual batch reports. This activity will require obtaining entity and facility IDs from EPA.

1.34: Please consider the following questions regarding situations where D&T may no longer be applicable:

a) How should deliveries across truck racks be reported? The trucking company or retail locations are not registered, so there will be no entity or facility ID. Should special codes be used to identify these?

    A: Designate and track ends at the point that fuel is dyed or marked (for those areas of the country where the fuel is subject to the marking requirement), or taxed- generally just before the fuel is given to a trucking company. Thus, they will not be registered, nor will they have ID numbers. There are still PTD requirements for these parties, and the PTDs associated with such hand-offs will denote both the transferor and transferee of the fuel, so special codes are not necessary. For reporting, truck racks must still track total fuel volumes dyed, taxed, and/or marked by grade. Truck terminals will also still be required to report deliveries by volume, sulfur level, designation, etc. (§ 80.600(b)).

b) D&T does not cover barrels of fuel after they leave the truck terminal before delivery to retail and does not deal with retail stations, correct?

    A: Designate and track ends at the point that fuel is dyed, marked, or taxes are assessed (for those areas of the country where the fuel is subject to the marking requirement). The dye and marker can then be used to ensure that fuels are not inappropriately shifted to other markets, and thus D&T is not necessary at that point. The addition of the dye and/or marker will generally occur at a point before the fuel exits the terminal, so D&T will not cover fuel after it leaves the terminal as long as it has been dyed, marked, or taxes assessed. Please note that the fuel will still be subject to the anti-downgrading provisions of § 80.527.

1.35: Can a truck terminal and the pipeline that brings product to the truck terminal be aggregated if that same truck terminal receives product from another common carrier pipeline?

A: *Yes, the truck terminal and pipeline may be considered an aggregated facility if they are both owned by the same party and if that party maintains custody of the fuel while it is within the boundaries of the aggregated facility. (Only refineries are precluded from aggregation in the case of a common-carrier pipeline.) For reporting, the aggregated facility would need to report volumes received into the pipeline as well as volumes received into the truck terminal from the common-carrier pipeline.*

1.36:   For a system with multiple truck loading terminals, terminal operators do not necessarily know which volumes are highway diesel and which volumes are nonroad diesel until the fuel is loaded. How can volumes be accurately reported at the truck loading terminal level?

     A: *The terminal operator is responsible for maintaining appropriate volume balances. A recordkeeping system that captures individual batch data is required for batches received into, as well as batches delivered out of, the terminal. Reports submitted to EPA summarize, by product type, receipt and delivery activities and results of compliance calculations for the applicable compliance period. A terminal operator would report subsequent to the loading/taxing/dyeing for that compliance period, so the fact that the designation is unknown until the point that it is loaded does not pose a problem (though, the individual deliveries must be tracked so that the terminal operator can accurately report at the end of the reporting period).*

1.37:   Can an entity register multiple facilities initially until a history or track record is established and then in a subsequent compliance period register fewer facilities by combining some?

     A: *Yes, facilities may change their facility boundaries and register some aggregated facilities in a subsequent compliance period (compliance periods are yearly), as long as the aggregated facilities do not violate the facility/aggregated facility provisions set forth in the regulations. However, a party's facility definition or boundary may not change within a given compliance period.*

1.38:   Do facilities for RFG and ULSD have to be the same? Will EPA assign new ID numbers for facilities that already have a facility ID under RFG? Will those existing facilities have to register again?

     A: *Facilities for RFG and ULSD can be the same, however they do not have to be the same. EPA may issue a new number if the facility definition changes. Entities should still contact EPA to update their registrations, even if you plan to use the same facility and facility definition that was used for RFG.*

1.39: Can non-contiguous "terminals for hire" that are owned and operated by the same party be aggregated? Also, can such terminals be aggregated if they receive product from common-carrier pipelines?

> *A: No. These terminals can only be aggregated if they are contiguous and if custody is maintained by the terminal at all times throughout the aggregated facility.*

1.40: a)   If a pipeline system does not transport 500 ppm HW diesel fuel and all NRLM is dyed before entering the system, would this pipeline be required to comply with the D&T requirements?

> *A: D&T ends completely at the point that fuel is dyed/taxes are assessed when fuel is being sold or sent off to retail for sale, the pipeline system would not have to comply with the actual designation and tracking of such fuels. However, in this case, the pipeline system would still have some of the D&T requirements, though the requirements are somewhat less. The pipeline would still need to follow the registration, reporting, and recordkeeping requirements associated with D&T, and it would still need to comply with the volume balance requirements.*
>
> *While the limited fuels you noted would mean there are perhaps fewer D&T requirements that apply, the pipeline system would still need to meet several designation requirements depending on the type of fuel that is delivered. For example, all 15 ppm highway or MVNRLM fuel has to be designated (§§ 80.598(b)(2) and (3)); and certain high sulfur fuel has to be designated (§§ 80.598(b)(5) and (6)). The pipeline is still subject to the volume balance requirements (e.g., anti-downgrading restrictions would apply if any 15 ppm highway needed to be downgraded to 500 ppm highway at the point of delivery). The fuel limitation noted above removes some, but not all, of the D&T requirements for the pipeline.*

b)   Similarly, if terminals along this pipeline system do not receive, store, or terminal any 500 ppm HW diesel fuel and all NRLM is already dyed when received, do these terminals have to comply with the D&T requirements?

> *A: Similar to part (a), above, there are fewer D&T requirements but the terminal would still need to meet some designation and volume balance requirements. For example, 15 ppm MVNRLM received and delivered by the terminal would still need to be designated, would be subject to the anti-downgrading provisions, and would still be subject to volume balance requirements.*

1.41: What is the D&T responsibility of a truck carrier that hauls off-spec product that has been neither taxed nor dyed back to a refinery or a transmix facility?

A: *The presumption in this case is that if the fuel is no longer classified as diesel fuel (and the PTD should reflect that), the fuel is no longer in the designate and track system.*

1.42:   A refinery produces a batch of diesel containing 10 ppm sulfur, and hands the batch off to a common-carrier pipeline at 10 ppm. The refinery wants to include the batch in its 15 ppm production volume, but the pipeline tells the refinery that the batch must be designated as 500 ppm diesel because the batch does not meet the pipeline's inlet sulfur specification. May the refinery include the batch as part of its 15 ppm production volume to demonstrate compliance with the 80/20 requirement for refiners and importers, but designate the batch as 500 ppm when it is handed off to the pipeline? If so, does the refinery or pipeline need to account for the batch in any downgrading calculations?

A: *Compliance with the 80/20 highway provisions is based on the designation of fuel produced. Per § 80.502(d), production of a batch is defined as "a quantity of diesel fuel or distillate...whose custody is transferred...to another facility." Therefore, a refiner's production is what is handed off to the next facility. While this fuel technically meets EPA's 15 ppm sulfur limit requirements, if a facility has an inlet sulfur specification, it is stating that the spec is the highest sulfur level that the facility believes that it can handle while still ensuring that the fuel will still meet the 15 ppm sulfur requirement when it is delivered to retail/the end user. Acceptance of a batch of fuel that exceeds a facility's inlet specifications could contribute to off-spec fuel being introduced into the system. (Please note that EPA is neither endorsing nor prohibiting the practice of an inlet specification being set; we recognize that some facilities may need to do so in order to ensure that a batch of fuel will ultimately satisfy the 15 ppm sulfur standards. Further, a facility that chooses to set an inlet specification may do so upon its own volition; a facility may also opt to take the added risk of accepting a batch of fuel that exceeds its sulfur specification if it so chooses.)*

Thus, the pipeline's request that the refiner designate the fuel as 500 ppm highway diesel at the point of delivery to the pipeline means that this batch of fuel would be treated as 500 ppm fuel for purposes of compliance with the refiner's 80/20 requirements. The pipeline would not be responsible for a downgrade; and in this situation, there would not actually be a downgrade since refiners are not subject to the downgrading restrictions (it would merely be reflected in the refiner's 80/20 requirements).

1.43:   Our company owns a refinery linked to company-owned terminal tanks outside the refinery via a pipeline that is also owned by the refinery. What are the requirements to be able to register the refinery and terminal separately, knowing that on a routine basis, refinery production is sent directly (actively) to the terminal tanks? Only fuel manufactured at our refinery is shipped on the pipeline. If the terminal and refinery could be registered separately, what would be the requirements to meet in terms of metering and analysis between the two facilities?

A: *The definition of facility starts with the assumption that a series of contiguous places where an entity maintains custody over diesel fuel is treated as a single aggregated facility. There are certain exceptions to this, for example a refinery may not be aggregated with a pipeline or terminal that receives fuel from other refineries. In addition, the entity generally can voluntarily subdivide this aggregated facility and register multiple facilities. In this case, assuming fuel from another refinery is not brought into either the pipeline or the terminal, then the combination of refinery, pipeline and terminal tanks could start off as one aggregated facility. The owner could also subdivide this and separately register more than one facility. For example, each component could be a separate facility, the refinery and pipeline could aggregate as one facility and the terminal as another, or the terminal could aggregate with the pipeline as one facility with the refinery as another. While an entity can change its choice in subdivision of facilities, this choice cannot be changed during a compliance period. Therefore, the owner does have the option to treat their terminal as a separate facility. (See § 80.502.)*

It is important to note that once the facility has been registered, it is treated as a separate part of the distribution system for purpose of D&T. All of the D&T requirements apply for any diesel fuel received or discharged from the separate facility, just as if it was owned and registered by another entity. It is also important to note that if a part of the system is aggregated with the terminal, such as the pipeline, then that entire aggregated facility is subject to the D&T requirements that apply to terminals. Likewise, if the pipeline is aggregated with the refinery, then that entire aggregated facility is subject to the refiner requirements. (See § 80.502(b)(1).)

In this case, the hand-off point for purposes of D&T would depend on the facility subdivision decided by the owner at registration. The decision on subdivision will then dictate the metering and analysis requirements that apply under D&T. For example, if the terminal is a separate facility, then the hand-off of custody from the pipeline to the terminal would be the point at which volumes and designations would be determined for purposes of D&T, for both the pipeline and the terminal. If the refinery and the pipeline are separate facilities, then the point at which custody is handed off from the refinery to the pipeline (the point of delivery) is the point at which volumes and designations would apply for purposes of D&T for both the refinery and the pipeline.

The refiner is responsible for certifying that the fuel meets EPA specifications and would need to conduct tests to support this certification. Facilities downstream of the refinery are not required to conduct tests to verify that the fuel meets EPA specifications. However, as one element in establishing a defense to presumptive liability should a violation be discovered, the regulations provide that downstream parties must have a quality assurance program that includes sampling and testing. If the refinery is aggregated with a downstream location, (e.g., in this case, if the refinery and pipeline are registered as one facility) then the aggregated facility is subject to the refinery requirements, and not to downstream D&T requirements. (See § 80.502(b)(1)(i).)

1.44: § 80.520(b)(2) states "*Until June 1, 2010, any #1D or #2D distillate fuel that does not show visible evidence of dye solvent red 164 shall be considered to be motor vehicle diesel...except for distillate fuel designated or classified as any of the following:...(vi) Diesel fuel that is produced beginning June 1, 2006, with a sulfur level less than or equal to 500 ppm, and designated as NRLM or LM that has not yet been distributed from a truck loading terminal or bulk terminal...*". Does this mean that after June 1, 2010 refineries must start dying the 500 LM, small refinery 500 ppm NRLM, and Credit 500 ppm NRLM? Will transmix processors be required to dye 500 ppm transmix derived NRLM/heating oil? Will parties upstream of the terminal be required to dye segregated interface derived 500 ppm NRLM/heating oil? Pipelines may be moving these products until 2014 and in-use generated LM 500 ppm indefinitely. Does the 15 ppm NRLM need to be dyed at the refinery at some point in the future or can it always be dyed at the terminal? Please clarify the June 1, 2010 date in the dye requirements section.

A: *The dye provisions in § 80.520(b) do not directly require that dye be added at the refinery gate. The dye provision establishes a presumption that fuel is highway fuel if it is not dyed. As a result, in the past refiners have added visible evidence of the dye to all fuel that did not meet the highway standard, so that it would not be treated as subject to the highway standard. In the 2004 nonroad rule, EPA dramatically limited the role that dyeing plays under the regulations, to allow for more fungible distribution of diesel fuel. After June 1, 2010, there is no longer any presumption that undyed fuel is highway fuel. Therefore after that date there is no need for refiners to dye fuel to make sure it is not considered highway fuel subject to the highway standards. At that point, EPA will rely on the designation provisions to identify what the fuel is and what sulfur standard applies.*

Prior to June 1, 2010, only a limited subset of diesel fuel must be dyed (in the distribution system above the truck rack; fuel must still be dyed when leaving the truck rack to meet IRS requirements) to avoid the presumption that it is highway fuel. From the effective date of the nonroad diesel final rule (August 30, 2004) until June 1, 2010, only heating oil and high sulfur NRLM must be dyed red at the refinery gate to avoid being subject to the highway and nonroad sulfur standards. Prior to August 30, 2004, all diesel fuel which did not meet the applicable requirements for highway diesel fuel was required to be dyed red at the refinery gate to avoid being subject to the highway standard. After June 1, 2010, there are no EPA requirements to dye any diesel fuel at the refinery gate. Similarly, there are no EPA requirements for diesel fuel produced downstream of the refinery such as transmix or segregated interface-derived diesel fuel to be dyed upstream of the terminal. In all cases, highway fuel is required to be free of visible evidence of red dye.

Any diesel fuel which is required to contain the fuel marker solvent yellow 124 (heating oil beginning 6/1/2006 and LM diesel fuel 6/1/2010 - 5/31/2012; see § 80.510) is also required to contain visible evidence of dye solvent red 124. The fuel marker and associated red dye requirements apply prior to the distribution from a truck loading terminal. The Internal Revenue Service requires that untaxed diesel fuel (nonroad, locomotive, & marine, & heating oil) be dyed before leaving the terminal. Thus, 15 ppm NRLM can be dyed just prior to leaving the terminal.

1.45: How will EPA deal with truck terminals located at a refinery? Are there any reporting obligations under D&T (i.e., quarterly balances, etc.)? Current Q&A question 2.39 addresses the sale of another refiner's product over its own truck rack, but what about the sale of a refiner's own product over that same truck rack?

A: *The manner in which the facility is registered will dictate reporting obligations of the truck terminal.*

If the truck terminal is registered separately from the refinery, then the refinery must meet all of the requirements applicable to a refinery, and the truck terminal must meet all of the D&T requirements applicable to a downstream truck terminal. The requirements of D&T- including hand-offs and balance calculations- must be reported, as appropriate, by the refinery and truck terminal.

If the truck terminal is aggregated with the refinery, then this single aggregated facility is treated as a refinery, and must meet all of the requirements applicable to a refinery. In this case, there would be no separate truck terminal D&T requirement for the portion of fuel being delivered over the rack, so long as requirements for dye/marker/taxes are satisfied. The hand-off from the truck rack would be treated in the same way as a hand-off from a refinery. The refinery's batch volumes and designations would be based on the hand-offs of fuel at the point it is released from the truck terminal, as that is the point at which the aggregated "refinery" would hand-off custody of the fuel. (See definition of "batch" at § 80.502(d).) In this special case, the refinery will have to provide batch reports showing volume of marked, dyed, and/or taxed product that was sold over the truck rack, as well as batch reports reflecting designation and delivery volumes for any diesel fuel handed off to a downstream party separate from the truck terminal, such as a pipeline. This will enable EPA to differentiate between fuel that was sold over the truck rack versus fuel that was sent through the D&T system.

20

1.46: Consider the following scenario- fuel enters a pipeline at, or below, 15 ppm. At some point in the pipeline, the fuel becomes off-spec. If this fuel is then delivered to a terminal at that off-spec level, but is then put into a tank with fuel that is below 15 ppm, and thus becomes 15 ppm (or less) again, how is this handled in D&T? Specifically, what do the pipeline and terminal have to report?

> *A:* *In this situation, the pipeline needs to redesignate the fuel at the point it is delivered to the next facility in the distribution system- the terminal, in this case. The designation of the fuel by the pipeline at the point of delivery needs to accurately reflect the sulfur level, therefore it will need to be redesignated as something other than 15 ppm. The volume and designation of the delivered fuel will need to be accounted for by the pipeline under its D&T requirements. For example, if the fuel is redesignated from 15 ppm highway to 500 ppm highway, this would need to be accounted for by the pipeline in showing compliance with the anti-downgrading requirements. The terminal would then report a receipt of 500 ppm fuel. The terminal may redesignate that volume as 15 ppm once it becomes 15 ppm fuel again. If the terminal 'upgrades' the fuel from 500 ppm NR to 15 ppm HW, and delivers it as 15 ppm fuel, then the terminal will need to account for this volume of 15 ppm fuel in showing compliance with its D&T requirements and volume balances.*

1.47: Is there anything in the regulatory requirements that would prevent a downstream blender from producing a batch of 15 ppm highway diesel fuel by mixing a batch of fuel designated as 500 ppm diesel fuel (but with an actual sulfur level only slightly over 15 ppm) with a batch of fuel designated as 15 ppm (but with an actual sulfur level considerably below 15 ppm)? Am I correct in that since no previously uncertified / undesignated blendstocks were used by the blender, that they would not be required to report as a refiner to EPA?

> *A:* *There is nothing in the regulatory requirements to prevent a downstream party from producing 15 ppm diesel fuel in the manner described. Since no previously uncertified/undesignated blendstocks were used, the party commingling/mixing the two fuel batches would not be required to report as a refiner to EPA. However, the sulfur level of the resulting fuel would need to be measured, and the fuel would need to be accurately designated based on its sulfur level. The designation and volume of the fuel as delivered would need to be reflected in the party's D&T requirements and volume balance calculations. If the 15 ppm and 500 ppm fuel used in the mixing operation described were both designated as highway diesel fuel when they were received, and the resulting fuel was designated as 15 ppm highway, there would be a neutral impact with respect to demonstrating compliance with the motor vehicle volume balance requirements (see section 80.599(b)). If the 500 ppm fuel used was designated as NRLM diesel fuel, there would be a negative impact with respect to demonstrating compliance with the motor vehicle volume balance requirements which would need to be*

21

accounted for by the party to show compliance during the same compliance period.

1.48: If a parent company has wholly owned pipeline and terminal subsidiaries, can the facilities be aggregated and reported by the parent if the facilities meet the contiguous facilities requirement? Can a parent report all facilities for its wholly owned subsidiaries or must facilities be reported by legal entity?

> A: *Each registered entity reports to EPA. If the parent company is the registered entity for each of its facilities, then the parent company will report. If each subsidiary facility is its own registered entity, then each of those entities must report.*
>
> *If the facilities are contiguous and custody is maintained by one entity (the parent owner in this case), they may be aggregated. In that case the owner must register as an entity, and the aggregated facility must also be registered (§ 80.597(c) and (d)).*

1.49: Will EPA require terminals to include language/provisions on compliance with the ULSD rules in their contracts with tenants? (*i.e.*, provisions that would have tenants confirm that they understand their obligations as custodians and that they will not change the designation of fuel stored in leased tanks.) Also, when terminals lease tanks, it would appear that the fuel would be entirely under the control of the tenants of the leased tanks. In enforcement, will EPA only look at the actions of the tenants, or will EPA take action against the parent terminals?

> A: *The diesel sulfur regulations do not require that terminals include language in contracts to lessees, however terminal owners may find it useful to include such language to ensure that lessees fully understand their obligations as tenants. Whether or not a terminal owner includes such language in its leases, as stated in the preamble, the terminal owner has full responsibility over leased tanks and must ensure that its total volume balance- of all tanks within the terminal- meets the required balance at the end of a reporting period (see § 80.601). Since the terminal owner is the responsible party, EPA reserves the right to take any appropriate enforcement action against the terminal owner. In addition, where the lessees have committed a prohibited act, for example by causing the terminal operator to commit a volume balance violation, or by distributing fuel that violates the sulfur standard, then EPA reserves the right to take any appropriate action against the lessees as well. As in all cases, EPA will evaluate each case individually to determine the appropriate enforcement response.*

1.50: a) Would EPA consider allowing a terminal to be subdivided into more than one facility for the purposes of registration under the D&T reporting and recordkeeping requirements when the terminal owner leases some of its assets to another entity? Under such an approach, the terminal owner and lessee(s) would have the option of obtaining separate registrations for the respective parts of the terminal that they control. Allowing industry this flexibility would be most consistent with the business relationship between terminal owners and their lessees. Unless the terminal owner and the lessee jointly agree that the lessee be covered under the terminal owner's registration, the terminal owner's responsibility should end once the fuel is delivered to the lessee's storage tank. There are currently no mechanisms by which terminal owners could receive the necessary information from their lessees or control their lessees actions.

A: *Due to concerns that allowing multiple registrations for a single terminal would greatly complicate EPA enforcement and compliance assurance activities, we specified that terminals may not be subdivided for registration purposes under the D&T system except under unique circumstances as approved by EPA (§ 80.502(b)(5)). In most cases the fuel goes through common piping/manifolds that are owned and operated by the terminal owner prior to entering the leased tankage and/or after exiting. Therefore, the terminal owner would still take custody of the fuel and be required to report. The leased assets are typically operated by employees of the terminal owner. Thus, it should not be unduly burdensome for terminal owners to obtain the necessary information from their lessees to comply with the reporting requirements under the D&T system. The terminal owner could use contractual means to ensure that it receives the necessary information from its lessees, and to ensure that its lessees do not cause a violation of EPA requirements (and to recoup potential damages should a violation occur).*

Based on the above discussion, we continue to believe that it is appropriate to preclude a terminal from being subdivided except under unique circumstances as approved/defined by EPA. One such unique circumstance where EPA believes it is appropriate to allow a terminal to be subdivided is when refining activities are occurring within the terminal by lessees, as discussed below. EPA will consider applications for the subdivision of a terminal, or other unique circumstances, on a case-by-case basis.

b)   Who should report in the following scenario? A terminal operator leases tankage to another company. The lessee uses the tankage for blending purposes. The lessee provides instructions to the terminal operator regarding tank blending, but has the tank certified independently and does not provide the operator information regarding the final product specification. Therefore, the terminal operator may not have the information necessary for accurate reporting. Would EPA allow for the subdivision of a terminal in this case?

A: *If the blending of previously undesignated blendstocks occurs in a terminal, such blending operations are considered to be refining activities. In such a case, one party must satisfy the refiner recordkeeping and reporting responsibilities. If neither party met the refiner requirements, then both the lessee and the terminal could be liable under the regulations for any refiner-related violation.*

Where a lessee of terminal assets performs fuel blending operations that make it a refiner, there are three different options that EPA believes would be appropriate for addressing both the refiner and distributor D&T responsibilities under the regulations. We have described these situations below.

In the first option, the entity responsible for the terminal would state in its registration form that it is also performing blending/refining activities and would take on refiner responsibilities for the lessee's blending operations, including the refiner responsibilities under designate and track. The terminal entity/facility would satisfy the obligations of compliance, reporting, and other D&T obligations for both its distribution activities and the refining activities of the lessee. The terminal entity/facility would then use contractual means to ensure compliance with the regulations. While the terminal entity/facility is taking on more burden in this option, EPA anticipates that there are a number of situations where this might be the preferred option.

In the second option, the lessee would register as a separate facility only for its refining operations (production of fuel from the blending of blendstocks), and would receive a facility identification for its refining facility. The entity responsible for the terminal would still be registered as a distributor and would still be treated as one single entity and facility for any volumes of previously designated fuel that the lessee merely receives and distributes as a distributor, for purposes of downstream D&T provisions (as in part (a), above).

The lessee, in its capacity as a refiner, would need to comply with the D&T requirements applicable to refiners, including appropriately designating all volumes of fuel that it produces by blending. The volume produced would only include the volumes produced from previously undesignated blendstocks. The lessee/refiner would be subject to all of the refiner requirements for the fuel it produces, including batch sampling and testing, and the 80/20 provision for highway diesel fuel. The lessee would also have to report to EPA on the volumes of fuel that it produces and distributes, including reporting the designation, volume, and the recipient of the fuel. In most cases the recipient would be the terminal entity/facility (whose manifolds and piping the fuel would go into). The terminal entity/facility would then report volumes received from the internal lessee/refiner (just as it would report volumes received from any other external facility). The lessee/refiner would have to maintain accurate PTDs for its deliveries of finished fuel to the terminal entity/facility.

Where a lessee/refiner blends previously designated diesel with undesignated blendstocks, the volume that is considered produced will just be the additional volume added by the blendstock. Further, the properties of the fuel (i.e., sulfur content) will be the sulfur content of the finished blend. If the finished blend changes the sulfur content of the previously designated fuel, this change would need to be reflected in the terminal entity/facility's D&T volume balance calculations and documentation. For example- if 1000 barrels of previously designated 15 ppm highway diesel were blended with 1000 barrels of higher sulfur fuel that was not previously certified (and the resultant volume of fuel now exceeds the 15 ppm standard), the entire 2000 barrels of blended fuel must now be classified as 500 ppm highway diesel fuel. The lessee/refiner would only count the newly produced fuel as 1000 barrels of 500 ppm highway diesel fuel in its refiner D&T documents, and this volume would be reported as fuel produced and delivered to the terminal entity/facility. The terminal entity/facility would then report a receipt, from the lessee/refiner, of 1000 barrels of 500 ppm highway diesel fuel. Since the remaining 1000 barrels of the blended volume was previously designated as 15 ppm highway diesel fuel, the terminal entity/facility would treat it as a downgrade from 15 ppm highway diesel fuel to 500 ppm highway diesel fuel (and there would be no hand-off from the lessee/refiner to the terminal entity/facility of this portion of the resultant fuel since it was previously designated before it was blended). The terminal entity/facility would need to account for this downgraded volume in showing compliance with the 20% downgrade limitation, assuming it was delivered from the terminal to another downstream party as 500 ppm highway diesel fuel.

The added complexity in situations such as this may prompt the desire to use the third option, discussed below, instead. However, this second option may be preferred in situations where the ability to measure hand-off volumes of previously designated diesel fuel between the terminal entity/facility and the lessee is difficult or impossible.

In the third option, the lessee would register as a completely separate facility. The lessee would check both "refining" and "distributing" on its registration form, and would be responsible for both its refining (production of fuel from blendstocks) and distribution (receiving and distributing previously designated fuels) operations. The terminal entity/facility would also be registered as its own facility. Assuming common manifolds and piping (owned by the terminal entity/facility) are coming into and out of the lessee's tankage, the lessee would report volumes received from and delivered to the terminal entity/facility under the designate and track regulations. As with the second option, the new lessee facility would be responsible for compliance with the fuel standards and reporting to EPA as a refiner for the fuel it produces from blendstocks. Unlike the second option, however, the lessee is also fully responsible for compliance with the D&T volume balance requirements and the anti-downgrading requirements as a distributor.

Anyone choosing options two or three must apply for approval. The most appropriate way to apply is at the time of registration. For option three, a letter should accompany the registration form, and should state the intent to register as a separate facility within another party's terminal/on another party's property. For EPA to allow the use of option two, a letter from each party involved will need to accompany the registration form indicating agreement to the responsibilities under option two. If registration is issued, it will also be considered approval for subdivision of a terminal (under § 80.502(b)(5)). Following registration, the lessee will be issued an ID number that consists of the lessee's entity ID and the terminal owner's facility ID.

As the details of situations where a lessee is performing blending operations may differ for each case, and are often complicated, we encourage anyone with this situation to talk with EPA prior to seeking approval for subdivision of a terminal.

c) Who should report if a terminal is owned by one company but operated by another? The operator may not provide the owner detail needed to meet the reporting requirements?

   A: The terminal owner could contract with the operator to report to EPA or could obtain the necessary information from the operator to report directly to EPA. In either case, the terminal owner would be the registered entity, and is the party responsible for compliance. However, in a situation where a terminal is completely operated by a lessee for the lessee's benefit only, we would allow the lessee to register the facility (or facilities) and it would be responsible for all associated recordkeeping and reporting.

2. Handling/Downgrading Issues

2.1:   As we understand it, the rule provides that interface/transmix may be sold as heating oil or high-sulfur NRLM from June 1, 2007 through May 31, 2010. During at least a portion of that time period, the sulfur standard for NRLM will be 500 ppm. Can the above-500 ppm transmix be blended into NRLM that has a sulfur content of 500 ppm or less? If so, what happens if the blending causes the sulfur content of the entire tank to exceed 500 ppm?

   A: Blending of greater than 500 ppm NRLM fuel with 500 ppm NRLM fuel is allowed. However, the entity performing the blending takes on the liability for the blend meeting its designation. If the resulting blend is greater than 500 ppm, then the entire batch of fuel must be designated as high sulfur diesel fuel. This option is limited in certain cases by the volume balance requirements, which is based on a comparison of the amount of high sulfur NRLM received to the amount of high sulfur NRLM delivered.

2.2:   If the two products mentioned in question 2.1, above, cannot be blended, do the regulations require the high-sulfur transmix to be loaded across a separate rack line and to be labeled differently on the bill of lading?

   A: As discussed in 2.1, in many cases they can be blended. However, if they are not, they would have different designations and require distinct product transfer documents.

2.3:   How will Minnesota B2 (2% biodiesel) be treated under D&T?

   A: Minnesota B2 is diesel fuel, and is therefore subject to all of the requirements of diesel fuel. Biodiesel must comply with the same sulfur specification applicable to diesel fuel from any other source. If sold as ULSD or blended with ULSD for later sale as ULSD, the biodiesel would be required to meet the 15 ppm sulfur specification applicable to ULSD. The terminal would be subject to presumptive liability if their addition of biodiesel to ULSD caused the 15 ppm sulfur standard to be exceeded.

2.4:   What is the terminal compliance date?

   A: The terminal compliance date for highway diesel fuel is September 1, 2006 (as amended by the November 22, 2005 technical amendments to the rulemaking) for all fuel they designate as 15 ppm diesel fuel.

2.5:   When do downgrade limitation dates start for pipelines?

A: *Anti-downgrade limitations begin October 15, 2006 for all entities in the distribution system.*

2.6: What requirements are there for clean trucks? Will trucks be dedicated to ULSD service? This will have major impact on where testing occurs.

> *A:* *There are no EPA requirements to use dedicated tank trucks to transport ULSD. It is the responsibility of the tank truck operator to ensure that whatever contamination that might take place while the ULSD is in the operator's custody does not cause the 15 ppm cap on sulfur content to be exceeded. Additional quality control measures may be needed to limit sulfur contamination if an operator chooses to use a tank truck compartment to alternately transport ULSD and high sulfur products.*

2.7: What limitations, if any, are there on the marketing of transmix that has jet or exempted product?

> *A:* *The special provisions applicable to the sale of products produced by a transmix processor using transmix as the feedstock are applicable regardless of the composition of the transmix. As a result of common pipeline batch sequencing, we anticipate that most transmix will contain some jet fuel and in certain areas heating oil. From June 1, 2006 through May 31, 2010, 500 ppm diesel fuel produced by processing transmix may be designated as highway diesel fuel and sold into the highway diesel market (40 CFR 80.513(a)). This flexibility applies to facilities that produce diesel fuel by processing transmix by distillation or other refining processes, but do not produce diesel fuel by processing crude oil. This flexibility only applies to the volume of diesel fuel produced by a transmix processor by processing transmix, and does not apply to any diesel fuel produced by the blending of blendstocks.*

2.8: Can a transmix processor produce 500 ppm diesel fuel for sale into the highway diesel market after June 1, 2006? May the transmix processor blend the fuel produced through the processing of transmix with a finished diesel fuel produced by a refiner (meeting a 15 ppm or 500 ppm sulfur cap) in order to ensure production of a final fuel blend that meets a 500 ppm sulfur cap for sale into the highway diesel pool? Without the ability to conduct such blending, the diesel fuel that transmix processors produce from transmix may not always meet a 500 ppm sulfur specification. This may limit the market for transmix-derived diesel fuel.

> *A:* *In the case described in the question, the transmix processor could designate and sell all of the 500 ppm fuel volume produced by blending processed transmix and 500 ppm diesel fuel into the highway diesel market, provided that the finished 500 ppm diesel fuel used for blending had already been certified by another refiner and designated as highway diesel fuel. If the finished diesel fuel used for*

blending was designated as No. 1 or No. 2 15 ppm highway diesel fuel, then all of the resulting 500 ppm diesel fuel could also be sold into the highway diesel market. However, the anti-downgrading requirements would apply to any No. 2 15 ppm highway diesel fuel used for blending. If the finished fuel used for blending was designated as something other than 500 ppm or 15 ppm highway diesel fuel, then the special provisions described above only apply to that fraction of the volume of the final fuel blend derived from transmix. For example: If 80,000 gallons of transmix was blended with 20,000 gallons of previously certified 500 ppm diesel fuel designated as nonroad, locomotive, and marine (NRLM) diesel fuel, then only 80,000 gallons of the finished fuel blend could be designated as 500 ppm highway diesel fuel. The remaining 20,000 gallons could be redesignated as NRLM diesel fuel (or heating oil). If the transmix was blended with a blendstock, and not previously designated fuel, then that fraction of the finished blend attributable to the blendstock would be subject to the standards applicable to refiners (e.g. 80/20).

We believe that the above discussion addresses the questioner's concern about the ability to manufacture 500 ppm diesel fuel from transmix. Furthermore, there will be outlets for diesel fuel produced by transmix processors that does not meet a 500 ppm sulfur specification during the time period when the highway diesel produced by transmix processors is subject to a 500 ppm sulfur specification. Such fuel may be sold into the NRLM market or into the heating oil market.

2.9:   Will a terminal have to wait for lab results before it can ship product to the rack or another pipeline?

   A: A terminal acting merely as a distributor or carrier is not required to wait for lab results before shipping diesel fuel. The regulations provide that, like other downstream parties, to establish a defense to presumptive liability a terminal must have a quality assurance program that includes sampling and testing, as one element of its defense to presumptive liability.

2.10: Very few truckstop operators have the ability to carry two grades of diesel fuel. The industry is not likely to install new tanks/lines/pumps for a four year phase-in period. Is there a minimum amount of each product that a marketer must carry and sell to qualify for the 20% downgrade exemption?

   A: If a retailer makes 15 ppm diesel fuel available to its customers in a manner consistent with the way it markets other fuels, then the retailer is not subject to the anti-downgrading requirements for highway diesel fuel. Typically, this would mean that the retailer would need to provide a fuel pump (or, pumps) and have sufficient volumes of 15 ppm highway diesel fuel available so that end-users wishing to purchase 15 ppm diesel fuel do not find it significantly more difficult to refuel than customers who wish to refuel with 500 ppm diesel fuel.

2.11:   At retail locations, will the diesel dispenser have a different size nozzle to prevent a customer from misfueling?

> *A: EPA did not require unique dispenser nozzles for 15 ppm diesel fuel. This is something we considered on several occasions, but concluded, after discussions with the relevant stakeholders that misfueling would not likely be significant enough to justify the cost and burden of the vehicle and fuel pump changes. The nonroad diesel rule includes labeling requirements for fuel dispensers and vehicles to help prevent misfueling. The fuel dispenser labeling requirements finalized in the highway diesel rule were superceded by those in the nonroad diesel rule (see 40 CFR 80.570, 80.571, 80.572, 80.573, & 80.574).*

2.12:   From an end-user viewpoint (trucking company):
a)   What is the liability in the event of misfueling by a driver?
b)   What effect does running ULSD in older equipment have on the engine?
c)   What is the impact of running 500 ppm fuel in a new (2007+) engine?
d)   Will companies with bulk fuel for use in their own equipment be subject to testing?
e)   Will existing tanks currently containing 500 ppm fuel be suitable for storage of ULSD?
-   will tanks need to be purged and cleaned
-   will company need to install new tanks to segregate fuels until 500 ppm is gone?

> *A: a)   See question 12.1. The trucking company is liable for misfueling but will be able to meet defense if it can demonstrate that the violation was not caused by the fleet operator or its employees (e.g., fueled from a retail pump labeled as containing the appropriate fuel for the vehicle) and, if it is from the company's own pump, it also needs to show PTDs account for the fuel and show it was compliant.*
>
> *b)   Generally, ULSD will be beneficial to engine operation and durability. The only known concern is with the lubricity of ULSD, which will be ensured through the use of lubricity additives to the fuel. ASTM has adopted a lubricity spec for diesel fuel to assure proper fuel lubricity.*
>
> *c)   If a new (MY 2007 or later) truck is misfueled <u>once</u>, it will have significantly higher PM emissions during operation on that fuel, but there should not be any significant long-term emissions or engine durability concerns as long as the vehicle is then fueled with the proper fuel. Constant misfueling would damage the aftertreatment/emission controls on these newer vehicles.*
>
> *d)   EPA will inspect wholesale purchaser-consumer facilities and take samples from fuel pump stands and from vehicle fuel propulsion tanks.*
>
> *e)   ULSD can be stored in tanks currently storing other fuels, including 500 ppm fuel and high sulfur fuel. However, care must be taken when*

transitioning the tank to avoid contamination of the ULSD. If a rapid turnover is desired, purging and cleaning the tank is an option, however it is not required.

2.13: It could take several loads of 15 ppm to flush out a retail tank, and this could take several weeks. Is there a provision for downgrading the product during this transition?

A: *Yes. During the start of the program, the anti-downgrading provisions do not start until October 15, 2006 (see § 80.527(c)(3)), allowing for a normal orderly transition. Even after October 15, there is an allowance for 20% to be downgraded. Since contamination at retail after the transition is expected to be minimal, 20% over the course of the entire compliance period should be sufficient.*

2.14: Regarding the 2 ppm adjustment factor- this applies downstream of the refinery and the import location. Assuming a terminal has a single ULSD storage tank that receives product from a refinery and also receives ULSD imports, both receipts would be designated as 15 ppm on-highway diesel and would otherwise be fungible. Would this tank be a downstream tank or an import tank? Would the 2 ppm adjustment factor apply to this tank or not?

A: *As in other fuels programs, tanks downstream from the certification tank will generally be considered "downstream" tanks and will be eligible for the 2 ppm test result adjustment, per § 80.580(d). However, this would not be the case if a company uses a tank as a downstream tank for refinery purposes and also uses it as a certification tank for its imports. In such a case the tank would not be eligible for the 2 ppm adjustment.*

2.15: What are the recordkeeping issues related to biodiesel used as lubricity additives for ULSD?

A: *Recordkeeping in this case will be the same as any other blending component. If the biodiesel provider designates the fuel as MVNRLM, then he serves as the refiner. If the provider does not designate, then the blender serves as the refiner.*

2.16: In order to be in compliance with the lubricity requirements of the ASTM D 975 diesel fuel specification, refineries will have to use lubricity additives, some of which may contain sulfur. Most, if not all, pipelines will likely prohibit the use of these additives in product that they transport. Therefore, these additives will be added to diesel fuel downstream of the pipelines (either into tanks at the terminal or at the rack). Can volumetric calculations be used to account for the sulfur content of these additives?

A: *No, volumetric calculations cannot be used to account for the sulfur content of these additives.*

2.17: During the initial phase, I deliver 15 ppm fuel to a customer. The product is contaminated (and, as a result, no longer meets the 15 ppm standard) while in his custody. The customer then ships the fuel back to me due to its contamination. Upon receipt of the fuel, I reclassify it as 500 ppm- does this count against my 20% downgrade?

> A: *The owner of the facility where the contamination occurs is responsible for redesignating off-spec fuel, and for accounting for it in its compliance calculations. The product that is shipped back to your facility must be appropriately designated and the product transfer document must reflect this designation. When the product is returned to you, it should already be designated as 500 ppm (since the party that caused the contamination is responsible for redesignating fuel that no longer meets the applicable standard), so you would not have to reclassify/redesignate the fuel and the downgrade would not count against your balance.*

2.18: 50,000 bbls of 15 ppm is discharged from an on-spec vessel. On the way to the tank, it became contaminated via pipeline. The receiving tank had 20,000 bbls in it, and the 50,000 bbls of contaminated fuel is put on top of the 20,000 bbls. What is charged against the terminal's 20% rule if it is redesignated as 500 ppm highway fuel?

> A: *Volumes are calculated based on receipts and deliveries for each custody holder. In this hypothetical situation, there would be 70,000 bbls of fuel designated as 15 ppm highway fuel received by the terminal, but 70,000 bbls of 500 ppm highway fuel delivered. (See 40 C.F.R. § 80.527(d).) This assumes that the 50,000 bbls of fuel was in the custody of the terminal when the contamination occurred in piping.*

2.19: Must diesel fuel treated as blendstock (DTAB) be off-spec, or can it be on-spec and "treated" as blendstock?

> A: *The DTAB provisions of § 80.512 may be used regardless of whether the diesel fuel is "off-spec."*

2.20: The 20 percent downgrade restriction in the highway rule is going to be extremely tight the second half of 2006 in large part due to refining production only being required 3 months in advance of retail sales and the first year being a partial year for the downgrade standard and, most critically, this is the first year for tank transition with a lot of higher sulfur distillate in the system. Many are planning on just in time delivery of ULSD from the refineries which means transportation will be faced with considerable downgrade during the transition period that limit downgrade to the on-road market to 20 percent. While no retailer is required to sell ULSD, anyone that handles ULSD may not downgrade more than 20% of the net volume received in any year into the on-road pool. This is likely to be a problem, especially for those moving small batch volumes. EPA needs to delay the start date of the downgrade limit period or get rid of it all together for this first year. Alternative, if all else fails, they should group the period with 2007.

> A: *As finalized in the nonroad rule, the start of the first compliance period for anti-downgrading was delayed to Oct 15, 2006 to allow for the expected downgrade during the initial transition to ULSD. The compliance period was also lengthened to end on May 31, 2007, allowing a longer period over which compliance can be averaged.*

2.21: If a pipeline does not have sufficient storage capacity for all grades of ULSD and exempted product, will the pipeline be allowed to blend at the terminal to prevent outages?

> A: *Yes, pipelines/terminals may blend different grades of fuel subject to the anti-downgrading limitations on highway ULSD and volume balance requirements for highway diesel fuel, nonroad diesel fuel, and high sulfur NRLM as applicable (see section 80.599).*

2.22: Please comment on the possibility of elimination of the red dye requirement for off-road diesel with the introduction of 15 ppm highway fuel and allowing 500 ppm "non-road" diesel to be shipped undyed as it leaves refining gate (to allow the distributors system to accommodate 3 grades of diesel)- please confirm that off-road diesel meeting the 500 ppm requirements may be commingled with 500 ppm on-road diesel? Dye removal, and the ability to commingle will greatly reduce the strains on limited tankage and will add needed flexibility, thereby reducing potential distribution and supply disruptions. Dyeing of "non-road" diesel would take place at terminal rack. A terminal could then handle 15 ppm highway, 500 ppm highway, and 500 ppm nonroad with two tanks common in terminals today. It would also help pipeline companies. Also, can 500 ppm off-road diesel be redesignated as 500 ppm on-road diesel? If so, are there any limits or special testing or documentation requirements?

> A: *While it does not eliminate the red dye requirement for nonroad diesel fuel, the nonroad diesel rule's designate and track provisions (D&T) allow diesel fuel with similar sulfur levels to be commingled and fungibly shipped up to the point of distribution from a terminal, where NRLM must then be dyed. 500 ppm diesel fuel designated as NRLM can be re-designated as 500 ppm highway diesel, but it is subject to the limitations stated in §§ 80.598 and 80.599.*

2.23: Could a fleet operator that owns and uses only pre-model year 2007 motor vehicles indefinitely use 500 ppm highway diesel fuel? Even after 2010? What if the fleet operator also produces his own motor vehicle diesel fuel?

> A: *The regulations at § 80.500(d)(4) state that beginning December 1, 2010 the sulfur content standard of § 80.520(c) (i.e., the 500 ppm sulfur standard) shall no*

longer apply to any motor vehicle diesel fuel. After, December 1, 2010 all motor vehicle diesel fuel must be 15 ppm or less. The regulations at § 80.530(b) state that after May 31, 2010, no refiner or importer may produce or import motor vehicle diesel fuel subject to the 500 ppm sulfur content standard.

2.24: A refiner certifies all of its kerosene as dual Jet/No.1 diesel at 15 ppm, and counts that volume as 15 ppm motor vehicle diesel fuel and accounts for it in the 80/20 TCO. If all the kerosene is then "downgraded" to Jet fuel, so none of it is available as 15 ppm motor vehicle diesel fuel, is the refiner still complying with the 80/20 rule even though none of the fuel is available as onroad diesel?

   A: *If a refinery designates kerosene as motor vehicle diesel fuel, then sells the kerosene as commercial jet fuel, then the reclassified volume of kerosene is not subject to any downgrading limitation, and does not impact compliance with the 80% 15 ppm diesel fuel production requirement for the Temporary Compliance Option.*

2.25: Terminal operators who import product from abroad are in most instances responsible for compliance with the highway diesel fuel rule. These importers understand that if a cargo of on-highway diesel fuel enters the U.S. and does not comply with the 15 ppm standard, they can choose to blend the product to specification if it is possible. These importers would then designate the product as compliant when it leaves the terminal gate. Would EPA please confirm this interpretation.

   A: *The nonroad diesel rule does contain a provision allowing importers to blend near-compliant diesel fuel containing slightly more than 15 ppm sulfur with diesel fuel containing less than 15 ppm sulfur to produce a compliant blend containing less than 15 ppm sulfur- this fuel may then be designated as "diesel fuel treated as blendstock", or DTAB (see § 80.512).*

2.26: If a terminal imports off-spec kerosene (higher than 15 ppm), terminal operators are planning to blend the kerosene with diesel fuel so that the ultimate diesel fuel leaving the terminal gate and entering into commerce meets the 15 ppm standard. Would EPA please confirm that this process is permissible for an importer?

   A: *The regulations at § 80.525(d) state that kerosene that a kerosene blender adds, or intends to add, to motor vehicle diesel fuel subject to the 15 ppm sulfur standard must meet the 15 ppm sulfur standard. Section 80.521(b) allows additives containing more than 15 ppm sulfur to be blended into motor vehicle diesel fuel at concentrations of one volume percent or less if the sulfur concentration of the resulting blend does not exceed the 15 ppm sulfur standard; however, kerosene is not an additive and therefore only the provisions of § 80.525 apply for kerosene blending (see also, 7.8 and 7.10, below).*

2.27: Refinery shipments by pipeline often occur with the producing company retaining title to a volume of product in the pipeline, while the pipeline actually has custody of the product. In this instance, do both the title holder and the custody holder have separate 20% downgrading potentials, or do they share a common 20% downgrading potential? How does one determine who accepts responsibility and reporting for a downgrade that may occur, such as pipeline interface, when the producing company may have title to the product, but the pipeline has custody?

> *A:* *As clarified in the nonroad rule regulations, the 20 percent downgrade limitation applies separately to each facility that has <u>custody</u> of the fuel when it is downgraded. The downgrade provision allows each party in the distribution system that is subject to it to downgrade a maximum of 20 percent of the highway diesel fuel for which it has custody on an annual basis. The downgrade limitation was incorporated to protect the availability of 15 ppm diesel fuel at the beginning of the highway diesel program. Actual contamination and downgrade is expected to be far less than 20%. However, 20% was selected to cover even the worst case scenario.*

2.28: In situations where custody changes to/from related entities (such as Company A refinery to Company A Pipeline to Company A Marketing), are these related entities each entitled to a 20% downgrade? Or, since they are related, are they viewed as one entity and therefore entitled to one overall 20% downgrade?

> *A:* *As defined in the regulations with the nonroad rule, a downgrade is accounted for on a facility-by-facility basis. Flexibility was included in the rule to allow aggregation of facilities at industry's discretion subject to certain restrictions (see § 80.502(b).*

2.29: A downgrade occurs only when the designation of a motor vehicle diesel fuel is changed from 15 ppm to 500 ppm. A change in designation to any other product, such as jet fuel, home heating oil, No. 4 diesel, is not a downgrade?

> *A:* *Correct. For the purposes of these requirements, "downgrade" refers to redesignating 15 ppm motor vehicle diesel fuel to 500 ppm motor vehicle diesel fuel. (See § 80.527(a)). Changing the designation of 15 ppm motor vehicle diesel fuel to any fuel that is not motor vehicle diesel fuel is not a downgrade, and is not volume limited.*

2.30: A "retailer" is limited to a single 20% downgrade of 15 ppm highway diesel fuel to 500 ppm highway diesel fuel unless the "retailer" offers for sale both 15 ppm and 500 ppm highway diesel fuels? Is the term "retailer" intended to apply on the individual service station or truck stop level, or does "retailer" potentially apply to a grouping of related retail outlets, such that if any one of the outlets offers 15 ppm diesel fuel, all outlets could freely downgrade 15 ppm highway diesel fuel to 500 ppm highway diesel fuel?

A: Since the downgrade provision was intended to protect availability of 15 ppm highway diesel fuel, a retailer or wholesale purchaser-consumer who sells, offers for sale, or dispenses either only 15 ppm motor vehicle diesel fuel or 15 ppm and 500 ppm simultaneously throughout the calender year is exempt from the 20 percent downgrading limit. A retailer or wholesale purchaser-consumer who does not sell, offer for sale, or dispense 15 ppm motor vehicle diesel fuel can downgrade 15 ppm diesel in an amount no greater than 20 percent of the total volume of motor vehicle diesel fuel it sells, offers for sale or dispenses annually. The limitation applies separately at each facility. Only contiguous facilities can be aggregated.

2.31:   We understand that during the period the temporary compliance option is in effect, a party will base compliance with the 20% downgrade requirement on the incoming and outgoing PTD records. Is the "incoming PTD record" reflective of the bbls input onto a pipeline or the bbls received at a terminal?

A: These provisions were clarified in the regulations finalized with the nonroad rule. Specific answers can be found at § 80.599 (e). Calculations are based in all cases on volumes received into the facility and volumes delivered.

2.32:   In the situation where a refiner owns and operates a proprietary pipeline, can 20% of the pipeline movements be downgraded as provided by § 80.527 (c)? If a third party operates the pipeline, can 20% of the pipeline movements be downgraded as provided in § 80.527(c)?

A: As long as the pipeline is registered as a separate facility, the anti-downgrading limitation applies to it regardless of ownership. If the proprietary pipeline were aggregated into the same facility as the refinery, then the anti-downgrading regulations would not longer apply. Rather, the refiner's production would be based on the volumes delivered from the pipeline.

2.33:   If a refiner annually ships 100 units of diesel (20 units of 500 ppm diesel, 80 units of 15 ppm diesel) and the pipeline moves 3 units to 500 ppm diesel as interface, the terminal receives 23 units of 500 ppm diesel and 77 units of 15 ppm diesel fuel. Is the terminal now allowed to downgrade 16 units (20% of 80 units) or 15.4 units (20% of 77 units)?

A: Each party in the distribution system is subject to the 20 percent downgrade limitation based on the amount of 15 ppm fuel for which it has custody on an annual basis. Therefore, if the terminal only receives 77 units of 15 ppm sulfur fuel. It may downgrade up to 20 percent of the 77 units of 15 ppm fuel for which it has received custody to 500 ppm highway diesel fuel.

2.34:   Is there some limitation for how long or when a retailer must sell the 15 ppm highway diesel before they would fall under § 80.527(e)(2) as opposed to § 80.527(e)(1)?

A: *Retailers or wholesale purchaser-consumers who sell only 500 ppm motor vehicle diesel fuel are subject to a downgrade limitation such that a maximum of 20 percent of the total volume of motor vehicle diesel fuel that they sell on an annual basis can be composed of downgraded 15 ppm fuel. Under § 80.527(e)(1), retailers or wholesale purchaser-consumers who sell, offer for sale, or dispense 15 ppm motor vehicle diesel fuel are exempt from the 20 percent downgrade restriction. This includes parties who sell only 15 ppm, or sell both 15 ppm and 500 ppm simultaneously.*

    To best promote the purpose of the downgrade requirements, which is to account for contamination without interfering with widespread availability of 15 ppm highway diesel fuel, the exemption only applies to retailers or wholesale purchase-consumers that sell, offer for sale, or dispense 15 ppm fuel continuously throughout the year beginning January 1 (or for 2006, October 15). "Continuously" here is meant to include normal business practices, including short shut downs for repairs, or delays in sales based on delivery problems. If at any time during an annual compliance period (calendar year) a retailer or wholesale purchaser-consumer ceases to sell 15 ppm, he/she would be subject to the 20 percent downgrade restriction of § 80.527(c) over the remaining portion of the compliance period.

2.35:   § 80.527 restricts the flexibility of all parties in the 15 ppm highway diesel fuel distribution system to downgrade 15 ppm highway diesel fuel to other classifications. Is it correct that retailers can choose to downgrade all, or a portion, of the 15 ppm diesel fuel delivered, provided it is properly labeled and sold?

    A: *The downgrade limitation was incorporated into the highway diesel program to protect the availability of 15 ppm highway diesel fuel at the beginning of the program. If distributors and retailers purchased 15 ppm highway diesel fuel in large quantities and sold it as, or mixed it with 500 ppm fuel, assurance of the availability of 15 ppm highway diesel fuel would be compromised. If retailers are selling only 15 ppm highway diesel fuel, they are by definition not selling 500 ppm fuel and the provisions do not apply. If retailers are simultaneously selling both 15 ppm and 500 ppm highway diesel, they are still satisfying the need for 15 ppm availability and thus, we chose not to apply the downgrade limitation to them as well. Only retailers who are not selling any 15 ppm highway diesel fuel (that is they are only selling 500 ppm fuel) are subject to the downgrade provisions such that of the total volume of motor vehicle diesel fuel that they sell in a year, only 20 percent of it may come from 15 ppm supplies.*

2.36:   The rule states that 15 ppm highway diesel fuel found in violation of the 15 ppm standard will be credited toward the 20% downgrade volume allowance. Could this fuel be downgraded to off-road usage and not count toward the 20% downgrade allowance?

A: *If fuel originally designated as 15 ppm motor vehicle diesel fuel subsequently does not meet the 15 ppm standard, it can be either: 1) redesignated as 500 ppm and counted toward the 20 percent downgrade limit for 500 ppm motor vehicle diesel fuel, or 2) redesignated as NRLM diesel fuel (in unlimited amounts).*

2.37:   Could 500 ppm highway diesel fuel in excess of the 20% allowable 15 ppm highway diesel fuel downgrade be shipped back to a refiner and avoid a determination that the downgrading entity is out of compliance?

A: *Compliance is on a facility-by-facility basis. The regulated parties must have redesignated the excess downgraded 15 ppm highway diesel fuel as something other than 500 ppm highway diesel fuel to avoid exceeding the 20% downgrading limit. If it is downgraded and redesignated as 500 ppm highway diesel fuel, then it is a downgrade, even if it is shipped back to the refiner. If it is not re-designated as 500 ppm highway fuel, then it would not count against the downgrade limit. Compliance with the downgrading requirements is on an annual calendar year basis. Any regulated party downstream of the refinery gate may downgrade more or less than 20% of any batch as long as compliance with the 20% downgrade limit is met at the end of the year.*

2.38:   To allow for the learning curve necessary in a period of transition, would EPA consider a delay in implementing the 20% downgrade rule for pipelines and terminals during the time the distribution system is being converted to 15 ppm diesel fuel, perhaps up until October 1, 2006, recognizing that distributors will have every incentive to minimize the downgrade of 15 ppm diesel fuel given the anticipated product price differential?

A: *The regulations at § 80.527(c)(3) have been amended (per the November 22, 2005 technical amendment to the regulations). The provision now states that the anti-downgrading provisions begin on October 15, 2006.*

2.39:   Can a small refiner who must produce 100% 15 ppm motor vehicle diesel fuel sell downgraded 500 ppm fuel at its refining truck loading rack?

A: *The nonroad rule modified these provisions to define a refiner's production as the volume delivered to the next entity. To allow for downgrade in the handoff, the production requirement was modified to 95 percent. (This is discussed in the preamble, but was inadvertently left out of the regulations. This error was corrected in a technical amendment to the regulations (which can be found on EPA's web site at: http://www.epa.gov/otaq/regs/fuels/diesel/diesel.htm#regs, July 7, 2005).) Any fuel produced and certified by another refiner can always be brought in and sold over a refiner's rack. The refiner in this situation must also report and maintain records as a fuel distributor under the designate and track regulations.*

2.40: There has been discussion on the "downgrading" provisions. What about upgrading? For example can a terminal operator, marketer or third party earn and sell credits for "upgrading" V500 to V15? We assume that these do not employ any fractionators, and are not "refiners".

  A: Only refiners and importers can generate credits per the regulations at § 80.531.

2.41: If both 15 ppm and 500 ppm on-road diesel are both available at a terminal [facility selling wholesale by the transport] is the terminal restricted to downgrading no more than 20% of its 15 ppm onroad diesel to 500 ppm onroad diesel?

  A: Yes. The regulations at § 80.527(c) state that persons who sell, offer for sale, dispense, supply, store or transport diesel fuel may not downgrade a total of more than 20% of the motor vehicle diesel fuel (by volume) that is subject to the 15 ppm sulfur standard of § 80.520(a)(1) while such person has custody of such fuel. Calculations to demonstrate compliance are found in § 80.599 (e).

2.42: In doing my quality assurance check, I find that my supplier has given me 15 ppm because he was out of 500 ppm. Can I upgrade the product that the PTD's say is 500 to 15?

  A: Yes.

2.43: Consider the following issue - as a batch of fuel moves down the pipeline, volumes are stripped off of the heart of the batch, since the interface at best stays the same or more likely will grow in length as the batch moves down the pipeline, the volume of material that must be downgraded at the end of the pipeline as a percent of the remaining batch volume will be much larger than at facilities upstream on the pipeline. The concern is that facilities at the end of the pipeline may have more difficulty in coping with the 20% limit on down grading, on a facility-by-facility basis.

  A: The 20% downgrade limitation applies separately to each facility in the distribution system. Facilities at the end of the distribution system only need to comply based on the volumes of 15 ppm fuel that they receive. Any downgrade that occurs upstream does not enter into their compliance calculations.

2.44: Terminals have no financial incentive to downgrade product from 15 ppm to 500 ppm. However, there is concern about contamination and the 20 percent downgrade limitation, particularly within the first several months following implementation. Will EPA issue a technical amendment that would allow the industry to downgrade the 15 ppm product without limitation, or at least remove the limitation for the first six months of the program.

A: EPA believes we have allowed terminals adequate lead time to identify and correct potential sources of contamination for 15 ppm diesel. The 20% downgrade limitation is on an annual basis, so if a terminal has to downgrade more than 20% of its 15 ppm diesel at the beginning of the program, it has the opportunity to compensate for downgrades greater than 20% by downgrading less of its 15 ppm highway diesel later in the year. Further, the anti-downgrading provisions do not begin until October 15, 2006, in order to allow for a normal transition at the start of the program.

2.45:   Question 2.35 deals with restrictions on the flexibility of all parties to downgrade 15 ppm motor vehicle diesel fuel to other classifications. If a small terminal has its supply of 15 ppm fuel contaminated with 500 ppm fuel, how much can the marketer use of that contaminated fuel that the terminal has downgraded to 500 ppm? Our interpretation of the rule is that the marketer handling 500 ppm can get only 20% of his 500 ppm supply from downgraded 15 ppm. Is this correct?

A: The regulations at § 80.527(c)(3) state that the 20% downgrading limitation shall be on an annual compliance period basis (however, the first and last compliance periods listed in § 80.527(c)(3) are slightly less than a year due to the June 1 program start dates). The terminal may sell all of the contaminated batch as 500 ppm fuel provided that the terminal's total annual downgrade percentage for the compliance period does not exceed 20%. This fuel would now be designated a 500 ppm by the terminal. As such, anti-downgrading provisions would no longer apply to it for any marketers that received it.

2.46:   If diesel fuel that is designated as 15 ppm fuel at a refinery is moved from the tank in which the fuel was sampled and tested to a "shipping tank" within the refinery gates, or to a tank that is outside the refinery gates but that is aggregated with the refinery, does the 2 ppm test adjustment under § 80.580(d) apply to diesel fuel? In other words, can such fuel be designated by the refinery as 15 ppm sulfur content diesel fuel even though a sample taken from a shipping tank is tested by the refiner to contain 17 ppm sulfur?

A: No. Unlike other fuels programs under 40 C.F.R. Part 80, a batch of diesel fuel is not defined in terms of its volume and designations until it is transferred to the next facility. See 40 C.F.R. § 80.502(d). Therefore, the 2 ppm downstream test adjustment cannot be applied until the fuel is in the custody of a facility downstream of the refinery or import facility.

2.47:   Due to batch sequencing in pipelines, some amount of "pipeline interface material" is generally created while fuels are being transported by a pipeline. Depending on a pipeline's practices (and, possibly the practices of its receiving terminal(s), as well), some amount of slightly off-spec material may be knowingly transferred to receiving terminals. Can EPA provide some guidance or recommendations on how to account for this interface material?

A: EPA understands that the pipeline industry has generally used certain ticketing and handling procedures based on long-standing practices and agreements with their shippers and those that receive fuel from them. (EPA has also been made aware of the fact that these practices and procedures are included in pipeline tariff Rules and Regulations filed with the Federal Energy Regulatory Commission.) EPA would like for parties to retain as much flexibility as possible in these situations, while still ensuring product integrity. It is our understanding that in these situations, a number of "cuts" and ticketing options can be made to protect fuel quality and ensure that receiving parties are made aware of the type of fuel that they will be receiving.

In situations where a pipeline makes a cut that involves handing off a portion of potentially off-spec fuel to the next facility, we would allow one of the following options:
<u>Additional language on a PTD</u>- a pipeline may add language somewhere on its PTDs denoting that a certain volume of the fuel being transferred is interface, such as "interface ULSD" (or some similar language). Use of language such as this should state the approximate volume of fuel that meets the ULSD requirements as well as the volume that is believed to be off-spec. Any downgrading/re-grading would be the responsibility of the receiving facility.
<u>Interface ticketing as a specific product</u>- a pipeline may create an "interface ticket" of the volume of fuel that the pipeline believes may be off-spec. With this option, two tickets will be printed- one for the volume of true ULSD and one for the volume of the interface product. Depending on contracts between parties, a pipeline may designate the fuel as interface ULSD (in which case, if any downgrading/re-grading were necessary, it would be the responsibility of the receiving facility) or the pipeline may re-grade the fuel itself (and thus, any downgrading/re-grading will be the pipeline's responsibility). The receiving facility may then decide how to handle this volume of interface ULSD.

2.48:   The regulations make reference to "distributors" in the regulations on anti-downgrading at § 80.527. As a jobber (tank truck delivery service), I am not sure if I am considered a distributor. And if so, is a jobber who delivers 15 ppm highway diesel fuel to retail stations that sell 15 ppm highway diesel fuel subject to the anti-downgrade requirements?

A: *Yes, a jobber is considered a distributor in this case– the fact that a jobber delivers fuel to retail facilities that sell 15 ppm highway diesel fuel does not excuse the jobber from the anti-downgrading requirements. It was suggested that if a jobber delivers only to stations that sell 15 ppm highway diesel fuel, the exemption from the anti-downgrading requirements that applies individually to such stations should also apply to the jobber. This might be a reasonable approach if it could be verified that such a jobber did not also deliver diesel fuel to facilities that handle only 500 ppm highway diesel fuel. However, the only way for EPA to be adequately assured of this would be to extend the full designate and*

41

track reporting requirements down through the retail facility. We believe that the burden associated with such an extension would not be justified, therefore jobbers are subject to the anti-downgrading provisions at § 80.527.

2.49: Please clarify how the downgrading restrictions for 15 ppm highway diesel fuel apply to retail facilities and the tank truck operators that deliver fuel to retail facilities. If a diesel retailer operates trucks to deliver fuel from a terminal to his retail facility, can the retail facility and the trucks be grouped together for determination of compliance with the downgrade limitations? If a retail outlet uses up its 20% downgrade allowance, can it use the surplus downgrading allowance from another retail outlet or from a tank truck operator? Do the downgrade limitations apply differently to private versus common carriers (tank truck operators)?

> A: *Each retail facility (e.g. truck stop, service station, convenience store, or card lock) must comply with the downgrade limitations for 15 ppm highway diesel fuel separately. Section 80.527(a)(2) states that the anti-downgrading limitation applies separately to each facility as defined under section 80.502 where there is custody of the fuel when it is downgraded. Section 80.502(b)(3) states that a retail outlet may not be aggregated with any other facility. Thus, a retail outlet may not be grouped with a tank truck for purposes of determining compliance with the anti-downgrade requirements. These requirements do not differ depending on whether or not the tank truck and retail outlet are owned by the same entity (i.e. whether the tank truck is a common or private carrier). Furthermore, the unused downgrade allowance of one facility may not be used bay another facility.*
>
> *For most types of facilities in the fuel distribution system, the anti-downgrade regulations require that no more than 20% of the 15 ppm highway diesel fuel that a facility receives may be downgraded to 500 ppm highway diesel fuel. There are no restrictions on the amount of 15 ppm highway diesel fuel that may be regarded to nonroad, locomotive, or marine diesel fuel, or to heating oil.*
>
> *Section 80.527(e) contains special provisions for retail outlets regarding compliance with the downgrading limitations. A retail outlet that sells only 15 ppm highway diesel fuel, or simultaneously sells both 15 ppm and 500 ppm highway diesel fuel throughout the calendar year is exempt from the downgrade limitation. Retail outlets that sell 500 ppm highway diesel fuel but not 15 ppm highway diesel fuel are subject to the anti-downgrade requirements. However, compliance with these requirements is determined differently than for facilities upstream in the distribution system. For those retail outlets that sell 500 ppm highway diesel fuel but not 15 ppm highway diesel fuel, up to 20% of the total volume of 500 ppm highway diesel fuel that they sell during an annual compliance period may have been 15 ppm highway diesel fuel that the facility received as 15 ppm highway diesel fuel and downgraded to 500 ppm highway*

diesel fuel. The current regulatory text at § 80.527(e)(2) was amended by the May 1, 2006 technical amendments to the regulations (see Section 16 for links to all technical amendments) to reflect the above discussion on how the downgrade limitation is applied to such retail facilities.

2.50: Do the downgrading provisions of § 80.527 apply to fleets? There are references to "retailers" and "wholesale purchaser-consumers", but I am not sure if fleets are included in either of those classifications.

> A: *A wholesale purchaser-consumer is a fleet. Per the regulations at 80.2(o), a wholesale purchaser-consumer is "any...ultimate consumer of...diesel fuel...which purchases or obtains...diesel fuel...from a supplier for use in motor vehicles or nonroad engines, including locomotive engines or marine engines and, in the case of...diesel fuel...receives delivery of that product into a storage tank of at least 550-gallon capacity substantially under the control of that person."*

3. Analytics R&D Issues/Test Methods

3.1: What rounding convention will be used for sulfur content tested by EPA?

    A: The sulfur specification is 15 ppm. Standard rounding conventions apply. For example, a measurement of 15.49 would be rounded down to 15 ppm.

3.2: The regulations at section 80.580(c)(1) include the following: "...*is correlated with the appropriate method specified in paragraph (a)(2) of this section.*" Is "(a)(2)" incorrect? Should it be "(b)(1)" which references ASTM D 6428-99?

    A: Yes, the current citation in the regulations is incorrect. It should be "(b)(1)." This error was corrected in a technical amendment to the regulations (which can be found on EPA's web site at: http://www.epa.gov/otaq/regs/fuels/diesel/diesel.htm#regs, July 7, 2005).

3.3: The regulations at section 80.580(c)(1) include the following: "...*is correlated with the appropriate method specified in paragraph (a)(2)(ii) of this section.*" Is "(a)(2)(ii)" incorrect? Should it be "(b)(2)" which references ASTM D 2622-03?

    A: Yes, the current citation in the regulations is incorrect. It should be "(b)(2)." This error was corrected in a technical amendment to the regulations (which can be found on EPA's web site at: http://www.epa.gov/otaq/regs/fuels/diesel/diesel.htm#regs, July 7, 2005).

3.4: ASTM and ISO frequently update their test methods; must an EPA approved method be resubmitted to the EPA for approval if the modifications to the test methods (that are made by ASTM and ISO) are minor and have no significant impact on the accuracy and precision of the method?

    A. All EPA approvals are specific not only to a given test method and laboratory, but also to a particular version of that test method. Thus, if ASTM or ISO released a revised version of a test method, any EPA approvals for previous versions of that same test method would remain valid and laboratories could continue to use the previous version. If the laboratory intended to use the revised version of the test method, formal approval must be sought from EPA. EPA is exploring procedures in the context of the current regulations under which a revision to a specific test method could be deemed inconsequential to its accuracy and precision. If such a procedure can be developed, EPA may be able to permit the qualification for the old version of a test method to apply also to the new version of that test method.

3.5:   Must a third party independent lab be "registered" with the EPA to get test methods approved?

      A: *Under §§ 80.584 and 80.585, each lab that will be testing fuel for compliance with the diesel regulations must obtain EPA approval on a laboratory-specific basis. The information required for approval is somewhat different depending on whether the test method is one that has been approved by a voluntary consensus-based standards body. However, there is no independent laboratory registration per se, as there is under the RFG program under § 80.65.*

3.6:   Are the EPA approved test methods for diesel automatically approved for use on kerosene or heating oils, or must the test method be approved for each product?

      A: *EPA does not require testing on heating oil. Consequently, there is no need to seek approval from EPA regarding the method used for the testing of heating oil properties. Any testing on #1 diesel fuel (kerosene) used in highway diesel or diesel nonroad, locomotive, or marine engines required by EPA must be conducted using an approved test method. The test methods approved by EPA for testing of the properties of #2 diesel fuel are also applicable for the testing of #1 diesel fuel.*

3.7:   Under the performance-based test method approach adopted in the non-road diesel sulfur rule, there is no designated sulfur test method as has been specified in previous regulations. Methods developed by consensus bodies, as well as methods not yet approved by a consensus body, qualify for approval provided they meet the specified performance criteria, as well as the recordkeeping and reporting requirements for quality control purposes. The rule specifies the precision and accuracy criteria that a laboratory must demonstrate in order to qualify its particular method of choice. API believes that the EPA enforcement lab should be required to meet or exceed the precision and accuracy criteria imposed on industry and that this should be a prerequisite to pursuing any enforcement action on a fuel parameter. Does the EPA Ann Arbor laboratory also need to demonstrate that its sulfur method(s) of choice meets the precision and accuracy criteria in the rule and maintain records for quality control purposes?

      A: *The EPA lab in Ann Arbor currently performs testing of fuel samples for enforcement purposes. EPA fully anticipates that its laboratory will have the capability to run several different test methods meeting the precision and accuracy criteria specified in the rule. However, whether we meet these criteria or not, is not a "prerequisite" to pursuing any enforcement action. Per § 80.611, evidence can be used in assessing whether a violation has occurred. Uncertainty in our own test results is just one of the factors that would go into any decision to take enforcement action.*

3.8:   Will EPA test terminals or retail outlets?

A: *EPA conducts sampling and testing of fuel at all points in the distribution system.*

3.9:   A batch of diesel is produced, sampled and tested on the last day of the reporting period but it is not shipped for several days until the next reporting period. With gasoline, the production date is used as the reporting basis. Should the diesel batch be reported as part of the earlier compliance period or alternatively as the "production" date which is the date of shipment from the refinery?

A: *Under the ULSD regulations as finalized in the nonroad rule, a batch of diesel fuel is defined as homogenous product that has been transferred to the next facility in the distribution system. Consequently, the only batches of fuel that should be reported are those that are transferred during the compliance period to another facility.*

3.10:   Can early release ULSD (prior to June 2006) be certified using designated or alternative methods, or must a PBMS qualified method be used?

A: *If the refiner wishes to certify it as 15 ppm fuel and obtain early credits, it must follow the PBMS requirements. For ULSD, the use of the designated and alternative test methods expired on 12/27/2004, so the only way to measure 15 ppm sulfur diesel fuel is by using a method that meets the accuracy and precision criteria at 40 C.F.R. 80.584 and is approved under the procedures set forth at 40 C.F.R. 80.585.*

3.11:   Performance-based laboratory qualification misses the sampling piece of the equation. D4057 allows several methods for manual sampling of tanks, if the method is not specified then the results can be unrepeatable. Will the sampling method be specified by the analytical methods?

A: *No. The sampling method used under D4057 will be at the user's discretion.*

3.12:   Will EPA consider allowing the volume accounting reconciliation (VAR) approach with lubricity additives greater than 15 ppm sulfur?

A: *We have no intent to do so. The VAR approach was developed for the specific situation of conductivity improver additives where no additive option existed with less than 15 ppm sulfur and the additive was required to be added at the terminal.*

3.13:   The nonroad rule does not explicitly exclude on-line sulfur analyzers for certification purposes; and industry plans to utilize on-line analysis that meets EPA's rule for precision and accuracy. Can on-line sulfur analyzers be approved for certification purposes if they meet the precision and accuracy criteria stated in the regulatory requirements, and if not, why not? Section § 80.581 (c)(1) allows the use of on-line analyzer test methods that have been approved per section 80.580. However, sections §§ 80.580 and 80.585 do not

46

reference on-line analyzers; do these sections apply to on-line analyzer approval as well, or does EPA plan on issuing separate procedures for on-line analyzer approval?

A: *The intent of § 80.581(c)(1) was not to imply that on-line analyzers could be used to certify a batch of diesel fuel. § 80.581(c)(1) states that the diesel fuel batch's composite sample must be tested with a test method as described under the provisions of § 80.580. The purpose of the provision is to provide the flexibility to refiners to release a batch of diesel fuel prior to that batch's composited sample being tested with a test method that: 1) is described under § 80.580, 2) meets the accuracy and precision criteria described under § 80.584, and 3) is approved under § 80.585. The test method that meets these three criteria on the composited sample of the batch is the official test result. In order to allow for the release of the batch of fuel prior to the composited sample being tested, we believe it is necessary to require that on-line analyzers meet the accuracy and precision requirements of § 80.584 and are qualified under the process of § 80.585 (see also § 80.581(c)(1)).*

The Agency believes a regulation change would be required to allow on-line blending systems (including on-line analyzers) to certify batches of diesel fuel. However, before this issue could be addressed in a rulemaking, the Agency believes that, at the very least, the following concerns would need to be addressed: 1) Acceptable extent of intra-batch variability during the course of a blend; 2) Appropriate statistical demonstration of blending system's ability to closely track results from bench analyses of composite samples; 3) Nature of on-going statistical quality control of blending process needed to ensure that measurement quality is maintained; 4) Nature of recordkeeping and EPA access needed to permit audit oversight; and, 5) Size limits on batches.

3.14:   Section 80.581(c)(1) appears to require composite sampling/testing even though an approved on-line analyzer method is used during blending. If this is true, why must the on-line analyzer method be approved if the analyzer results are not used for compliance purposes? Will the composite testing and reporting requirement be waived if the on-line analyzers are approved? Will EPA consider on-line analyzer results as evidence in establishing a defense in enforcement actions?

A: *Yes, § 80.581(c)(1) does require that a composite sample of a diesel batch be taken, and the composite sample is required to be tested with a test method described under § 80.580, which meets the accuracy and precision criteria under § 80.584 and is approved under the process at § 80.585. But the intent of § 80.581(c)(1) was only to provide flexibility to refiners that utilize on-line analyzer equipment so they could release a batch of diesel fuel prior to its composited sample being tested by a test method described under § 80.580. In order for a refinery to have the flexibility to release a batch of diesel fuel prior to its composited sample being tested with a test method described under § 80.580, §*

80.581(c)(1) does require that the on-line analyzer must meet the applicable accuracy and precision under § 80.584 and be approved under the process at § 80.585. This is to help ensure that the diesel fuel leaving the refinery is compliant with the applicable standard. Thus, on-line analyzer approval is required if a refinery wishes the flexibility of releasing a batch of diesel fuel prior to that batch of diesel fuel's composited sample being tested by an approved test method under § 80.585. Thus, EPA will not consider on-line analyzer test results as meeting the requirement for refiners to test each batch of diesel fuel designated as 15 ppm fuel.

3.15: Given the results of the recent ASTM round robin testing comparing the repeatability and reproducibility of the designated ASTM 6428 sulfur test and the alternative ASTM 5453 test, will EPA reconsider the issue and define the more precise ASTM 5453 test method as the designated sulfur test method?

A: *The regulations finalized with the nonroad rule modified the sulfur test method requirements. Instead of a designated method, there are now simply performance-based requirements that must be met, allowing the use of multiple test methods (see § 80.580 and 69 FR 39184).*

3.16: Given the reproducibility of even the most precise sulfur test method is no better than 6 ppm, will EPA revise the 2 ppm enforcement test tolerance?

A: *Based on the results of EPA's 2005 ULSD Round Robin Test Program, we have temporarily changed the enforcement tolerance from 2 ppm to 3 ppm effective June 1, 2006 through October 14, 2008. On October 15, 2008 the enforcement tolerance will revert back to 2 ppm.*

3.17: If in general pipelines can demonstrate that the percentage of sulfur tracks with gravity, or sound velocity, or color, will that be a sufficient defense that the pipeline did not contaminate a batch (assuming the pipeline can show what readings were when the valve change were made)? Or will the pipeline need to have a lab analysis, go/no go test, or on-line sulfur analyzer to defend? The treatment of pipeline interface will be important to pipeline operations and the impact of 15 ppm diesel fuel implementation on fuel supplies.

A: *The rule does not specifically require that pipelines use on-line sulfur analyzers for quality assurance purposes. However, use of on-line analyzers may be appropriate at certain locations on the pipeline. To ensure that the diesel fuel meets the 15 ppm standard, the quality assurance program must include testing, and at this time EPA believes such testing must measure the actual sulfur content of the fuel. However, other relevant procedures that, for example, help a pipeline to locate an interface, could also be an important part of the quality assurance program for a pipeline.*

The rate of sampling and testing, and where product should be sampled, are a function of the circumstances attendant to a particular distribution system. The pipeline should use its knowledge of the system and exercise its best professional judgment in determining the rate of sampling and testing and where samples should be taken. Also, pipelines may act cooperatively with upstream and downstream parties to share their testing information as part of the pipeline's quality assurance program.

3.18: Due to the known reproducibility problems in testing at a 15 ppm level, would EPA consider letting downstream entities "average" the sulfur level tests within a range - say no more than 20 ppm maximum and 15 ppm average - to avoid supply disruptions?

     A: No. As stated in the preamble to the final rule, the sulfur sensitivity of emission controls that will be used on model year 2007 and later motor vehicles requires that the sulfur content of highway diesel fuel dispensed into 2007 and later heavy-duty vehicles not exceed 15 ppm. Consequently, under the rule, the 15 ppm sulfur standard is a cap that must be met on a per-gallon basis. However, during the transition period, highway diesel fuel of up to 22 ppm may be sold as ULSD (beginning June 1, 2006 through October 14, 2006).

              Also, to account for test variability downstream of the refinery gate or import facility, we allow for a downstream test adjustment of negative 3 ppm through October 14, 2008 (this adjustment will revert back to 2 ppm on October 15, 2008). The purpose of taking testing variability into account in compliance determinations for fuel sampled downstream of the refinery or import facility is merely to ensure that fuel actually meeting the 15 ppm cap is not rejected and treated as noncompliant due to concerns about testing variability. It is not expected to result in any increase in the actual sulfur content of highway diesel fuel above 15 ppm at any point in the distribution system. Note, however, that the rule does not limit the ability of the fuel distribution industry to set a commercial pipeline sulfur content specification. We acknowledge that pipelines may elect to set sulfur specifications.

3.19: If a pipeline or terminal has test results indicating that motor vehicle diesel fuel sulfur content is 15 ppm and EPA tests show the sulfur content is greater than 18 ppm, would the pipeline's tests be an acceptable defense, or would the EPA's test results prevail? This is of concern given the variability in test tolerance experienced during the round robin.

     A: EPA would treat this as a violation.

3.20: Pipelines require durable, easy-to-use test equipment that can quickly give test results. We have been fortunate to find equipment that meets our needs in time for past rule implementations. The EPA's field testing practice has been to use the same equipment pipelines use to conduct oversight. What are EPA's plans on field oversight in implementation of this rule, and has the EPA made any selection of equipment that will be used in the field?

A: *EPA does plan to conduct field screening of diesel fuel as well as testing diesel fuel using the designated method for enforcement purposes. We have not yet selected a field test instrument.*

3.21: ASTM updates test methods every five years. Will EPA issue a ruling that will allow the latest (most up to date) version?

A: *The nonroad diesel rule contains a performance-based provision specifically for diesel sulfur that allows the use of any diesel sulfur test method that meets certain criteria for accuracy and precision. Under this approach, industry would be able to determine on their own whether updated ASTM test methods would be acceptable for compliance testing purposes.*

3.22: Accuracy can be compared to a gravimetric standard? Are NIST standards available?

A: *Several highly accurate standard reference materials (SRMs) are now available from NIST.*

3.23: Is sulfur measurement at 15 ppm level practical at the pump? Do any of these methods work for portable/field testing?

A: *Currently only for sampling at the pump and testing in the lab. However, a portable MDWXRF instrument (ASTM method D 7039) is being considered for field testing use.*

3.24: Low sulfur gasoline test method is D-2622 [WDXRF], low sulfur diesel will be tested by D-6428 [5453 alternate]. Is there a chance that both test methods could be merged so that one test method for both low sulfur gasoline and diesel would be permitted?

A: *The Agency has adopted a performance based test method approach which would set forth criteria to determine the acceptability of use of voluntary consensus standard based analytical test methods. Under this approach, industry would be able to determine whether test methods allowed under the low sulfur gasoline rule may also be utilized under the low sulfur diesel rule.*

3.25: If 15 ppm is required in the retail market, what will be the pipeline requirement for sulfur at the refinery gate?

A: *EPA estimated that, on average, 7-8 ppm sulfur would be needed the refinery gate in order to produce 15 ppm sulfur at the pipeline's end (see the highway diesel rule Regulatory Impact Analysis, chapter V, section C.1.a (this chapter can be viewed on the web at http://www.epa.gov/otaq/regs/hd2007/frm/ria-v.pdf) for additional discussion). Sulfur contamination during pipeline transport will vary between distribution networks, and therefore refinery production sulfur levels can vary as well.*

3.26:   Terminal operators currently marketing ultra low-sulfur diesel fuel (15 ppm) are concerned about the accuracy of the equipment available today in the field for testing. Will EPA recommend which equipment industry should use for testing? What equipment has the Agency decided to use to determine compliance?

     A: *Presently we are utilizing several different instruments in our lab. We will use whatever instruments over time we believe provide the best result.*

3.27:   Is the Agency going to publish a list of sulfur analytical methods that had been deemed equivalent under the rule?

     A: *There is no designated method or equivalent methods for sulfur. We now have a performance based system where labs have to meet precision and accuracy criteria for whatever method they choose to use. A number of methods should be able to be used to meet the precision and accuracy criteria, but this will be done on a lab-specific basis.*

4. Regulatory Clarification and Thoughts/Questions for EPA

4.1: What, if any, grace or discretion will be afforded during start up of the enforcement process?

> A: *There are several provisions in the rule that provide flexibility during the transition period. For example, the 20% downgrade limitation is an annual compliance period requirement, such that any start up difficulties can be offset later in the year; and compliance with the downgrade limitations do not begin until after the transition period is over (October 15, 2006). Further, during this transition period, we allow for a 7 ppm adjustment, such that fuel up to 22 ppm can still be considered ULSD (until October 15, 2006). We also allow a downstream test adjustment on top of the 15 ppm cap- this downstream test adjustment is 3 ppm through October 14, 2008, and will become 2 ppm beginning October 15, 2008. Moreover, the rule provides substantial time periods for parties at each stage of the distribution system to come into compliance during the initial transition into the sulfur programs. Hardship provisions that are available to qualifying refiners and importers address cases of extreme hardship or unforeseen circumstances outside the control of the refiner or importer. However, under the "hardship" provisions of the regulations, under no circumstance may fuel having a sulfur content exceeding 15 ppm be sold or dispensed for use in engines requiring 15 ppm fuel.*

4.2: The regulations talk about the requirements for PTDs for both the diesel fuel and the additive, but it is unclear at what point an additive PTD is tied to a fuel PTD; i.e. you would never know in which batch a high sulfur additive was used. Are we required to re-certify 15 ppm diesel fuel downstream if an additive with more than 15 ppm sulfur is used downstream of the refinery certification? Are we required to modify the diesel PTD to include language stating that an additive with more than 15 ppm sulfur was used in the product?

> A: *There is no specific requirement to tie additive PTDs to diesel fuel PTDs. Sections 80.521 and 80.591 set forth requirements for additives and additive PTDs. The regulations also set forth prohibitions against selling diesel fuel represented to meet the 15 ppm sulfur standard unless it does meet that standard. The additive blender will be liable if it causes fuel it represents to meet the 15 ppm standard to exceed that standard. For an additive blender who uses an additive having a sulfur content exceeding 15 ppm to meet its defense elements it must, among other things, show that: it did not cause the violation, that product transfer documents account for the product (both the diesel fuel and the additive), and indicate the fuel and additive were in compliance while under the party's control. Also note that § 80.613(d)(2) will require, in most cases, that the blender show that it tested every batch of fuel it blends such additive into. To establish*

these defenses, the additive blender would likely need to have records that do tie the additive blending to particular transfers of fuel. In the case of static dissipater additives, a blender may use a volume reconciliation approach in lieu of every batch testing.

4.3:   Is there any economic hardship requirement escape clause for the distribution chain? Given no field testing equipment available, insufficient tankage, commingled manifolds.

    A: *The economic hardship provisions in the regulations only apply to refiners.*

4.4:   What recourse does a pipeline have that has received an off-spec batch of ULSD from another pipeline?

    A: *The PTD for any given batch of fuel in the distribution system must accurately reflect the sulfur content of that batch. If a pipeline accepts a batch of ULSD into its system and later finds that the batch no longer complies with the sulfur specification for ULSD, then that pipeline must ensure that the PTD that accompanies the fuel as it leaves the subject pipeline's custody is accurate. If the pipeline finds that the sulfur content of the subject fuel batch is less than 500 ppm, then the pipeline may redesignate that fuel batch as 500 ppm highway diesel fuel (until 6/1/2010) provided that the pipeline remains in compliance with the 20 % downgrade requirements for 15 ppm highway diesel fuel and highway diesel fuel volume balance requirements. The fuel batch may also be redesignated as NRLM, LM, or heating oil provided that sulfur content of the fuel batch is consistent with the applicable sulfur specification. There are no limitations on a downstream pipeline's ability to receive such a batch of fuel.*

4.5:   What are the provisions from the non-road rule that have been expanded to the highway diesel rule, or that amend the highway diesel rule?

    A: *The nonroad rule final regulations (69 FR 38958, June 29, 2004) identify all the provisions that were amended from the highway diesel rule. Provisions that were changed are specifically noted at the beginning of the applicable section. However, please feel free to contact EPA staff if you have questions on a specific section.*

4.6:   The regulations require batch reports for small refiners for each compliance period from June 1, 2007 through May 31, 2010 per § 80.604(d). For each batch, this includes "*the sulfur content and cetane and aromatics content of the fuel*" (§ 80.604(d)(5)). This is not consistent with the Agency's response to comments, May 2004, p. 10-32, 10.3.3.3 Aromatics Reporting Requirements.

    A: *The regulations require batch reports for <u>all</u> refiners. This provision was erroneously copied into the reporting provisions in the regulations. We agree* *that the batch reporting of cetane index or aromatics levels for off-highway diesel fuel is not necessary. The regulations regarding a minimum cetane index of 40 or a maximum aromatics content of 35 volume percent for highway diesel fuel do not include any batch reporting requirements. This is also the case for sulfur content where only the designation is required. We see no need to impose such batch reporting requirements on NRLM diesel fuel at this time; however, records retained must contain this information. This was corrected in a technical amendment (signed July 9, 2005 and published in the Federal Register on July 15, 2005), which can be found at: http://a257.g.akamaitech.net/7/257/2422/01jan20051800/edocket.access.gpo.gov/2005/pdf/.*

4.7: What provisions exist in the ULSD rules for importing nonroad and highway diesel fuel from Canada by truck?

    A: *Provisions for diesel fuel transported into the U.S. by truck or rail car are found in § 80.583. They are similar to the gasoline sulfur rule provisions for importing gasoline into the U.S. by truck.*

4.8: How would distillate not used as fuel (i.e. for freeze protection) be accounted for?

    A: *The possible designations for #1 and #2 distillate are detailed in § 80.598. An entity would have the option to designate it as heating oil or blendstock if it did not want to comply with the highway and NRLM standards.*

4.9: For terminal reporting, is it: quarterly reporting with quarterly compliance with 20% downgrade limitation, or quarterly reporting with annual compliance with 20% downgrade limitation?

    A: *Quarterly reporting and downgrading compliance are separate issues. All facilities in the D&T system have quarterly reporting requirements. Truck Loading Terminals also need to provide required compliance calculations. Downgrading calculations are a recordkeeping requirement, not a reporting requirement. The downgrading compliance periods are listed in § 80.527(c)(3).*

4.10: Current <500 pm kerosene blending into 500 highway diesel is an increase in the highway diesel pool. Is EPA revising § 80.525 to allow for this increase?

    A: *Section 80.525 contains the requirements applicable to kerosene blenders. The provisions regarding the determination of compliance with the volume balance for motor vehicle (highway) diesel fuel are contained in § 80.599(b). Each facility must maintain a positive or neutral highway diesel balance during any compliance period. The questioner is correct in that the addition of #1 500 ppm diesel fuel not designated as highway diesel fuel (such as 500 ppm kerosene) to*

*500 ppm highway diesel could result in an increase in the highway diesel pool.
This is possible, but would have to be accounted for in calculating compliance
with the facility's highway diesel volume balance. We anticipate that 15 ppm #1
diesel fuel will be available for wintertime blending. The use of such 15 ppm fuel
for wintertime blending into highway diesel fuel will not adversely impact the
facility's ability to comply with the highway diesel volume balance requirements.*

4.11: For consumers with a 2007 vehicle how will they be educated to look for sulfur content
at a station to prevent inadvertent use of 500 ppm?

A: *The fuel regulations require that all fuel pumps be labeled to inform the user
which fuel they are dispensing into their vehicle. These labels will state which
engines/vehicles that the specific fuel is suitable for use in. There will also be
labels on the 2007 engines, per § 86.007-35, stating which fuel should be used.*

4.12: Will it be legal for an Alaska refiner to dye its highway ULSD?

A: *The highway and diesel fuel regulations include no dye requirements or
restrictions for Alaskan fuel, provided the requirements of 40 C.F.R. §§ 69.51 and
69.52 are met. IRS has its own fuel tax requirements.*

4.13: Has EPA defined the quantitative amount of dye that is visibly dyed? And what does that
mean in a lower limit of dye?

A: *There is no set amount of red dye that must be added to fuel, EPA regulations do
not specify what concentration is required. The regulations only require that
"visible evidence" of red dye be present (§§ 80.510(d)(5),(e)(5), and (f)(5)), and
this amount may vary for different parties. This is the same visible evidence
criteria that was first implemented for non-highway diesel fuel in 1995 (§
80.520(b)).*

4.14: In an effort to maintain balance between nonroad and highway diesel fuel, EPA has
established rules pertaining to "shifts" in volume (anti-downgrading). Does this hinder
the establishment of new product distributors?

A: *We do not believe that the volume balance requirements will inhibit the
establishment of new product distributors. Any new fuel distributor need only
register and then comply for the portion of the compliance period in which they
were in business. New distributors would begin with a clean slate with respect to
demonstrating compliance with the volume balance requirements. In cases where
certain product distribution assets are sold to a new party in the middle of a
compliance period, EPA will provide guidance regarding how the previous and
new owners will work together to demonstrate compliance for the compliance
period in which the sale takes place. One option would be for the original owner*

to be responsible for demonstrating compliance during the portion of the compliance period during which it maintained ownership. The new owner would then be responsible for demonstrating compliance during the portion of the compliance period over which it exercised ownership. This would be EPA's default assumption regarding how the applicable requirements would be applied.

4.15: What is EPA's response to the distribution complexities outlined at the November 2004 workshop? Specifically, if the refiners are not producing ULSD until May 2006, what is EPA's belief that 15 ppm ULSD can be supplied at the pump by September 2006?

A: *We believe that the rule provides ample time for transition. In most cases, retail stations should have sufficient time to make the transition (which was extended to October 15, 2006). However, there is no requirement that retailers must sell 15 ppm ULSD. If their transition takes longer, then they may continue selling 500 ppm fuel longer.*

4.16: When may importers begin classifying distillate product as DTAB?

A: *On the effective date of the regulation.*

4.17: Will annual attestation be required for ULSD like it is for RFG today?

A: *There is no attestation requirement in the diesel fuel regulations.*

4.18: If the 80/20 provision allows for a 20% downgrade, can we assume that 100,000 bbl ULSD can result in:
- 80,000 bbl to the pipeline,
- 64,000 bbl at the terminal, and,
- 46,000 bbl at retail?

A: *The 80/20 (§ 80.530(a)(3)(i)(A)) and anti-downgrading (§ 80.527) are two separate provisions. The 80/20 provision is the requirement that at least 80% of a refiner's highway fuel produced is required to meet the 15 ppm standard. The anti-downgrading provision, which does not apply to refiners, allows for up to 20% of an entity/facility's 15 ppm highway diesel fuel to be downgraded to 500 ppm highway diesel fuel (QA 4.20 also discusses anti-downgrading).*

A refiner's production is defined as the amount of fuel that is actually delivered to the next party. The volume of 15 ppm highway diesel fuel delivered by a refiner to a pipeline will be reported to EPA based on the measured volume as received by the pipeline operator. In such a case, there would be no downgrade reported in moving the fuel from the refiner to the pipeline, as downgrading does not apply to refiners. Each custody holder of 15 ppm highway diesel fuel may downgrade up to 20% of the volume it receives to 500 ppm highway diesel fuel. There is no restriction on the volume of 15 ppm highway diesel which can be downgraded to products other than 500 ppm highway diesel fuel (e.g. heating oil, or 500 ppm NRLM diesel fuel). The 20% figure was set because we believed this would be the greatest percentage any one custody holder would need to downgrade due to the procedures necessary to limit sulfur contamination during the transportation of 15 ppm diesel fuel. We expect that most custody holders in the distribution chain will downgrade a much lower percentage of the 15 ppm highway diesel fuel they receive. In most cases economic factors will limit downgrading to the bare minimum. As a result, while the example is hypothetically possible, it is not a realistic scenario.

4.19: Has anyone given any thought as to what the other parameters will look like once ULSD is fully produced (i.e. aromatics, cetane number, nitrogen, etc.)? Are there any issues with respect to the energy content of diesel with the increased hydrotreating potential "lightening" of diesel going down to 15 pm?

A: *There is a projected change in density and energy content per gallon, thus leading to a small change in fuel economy. In most cases it should be imperceptible- since the change still falls within the range of normal variations in diesel fuel. While fuel economy decreases slightly, there is not an overall energy loss, since refinery production volume increases to offset the energy density change. There is also expected to be a slight improvement in the cetane number of the diesel fuel resulting from a small decrease in aromatic content due to hydrotreating.*

4.20: Please explain 20% downgrade provision in detail and specificity- how does it work at each stage in the distribution system, especially if one company plays many roles?

A: *The anti-downgrading provision is to prevent the intentional commingling of 15 ppm highway diesel fuel and 500 ppm highway diesel fuel that would result in no availability of 15 ppm highway diesel fuel. The 20% limit is there to allow for unintentional mixing/normal contamination during the compliance period. It works the same throughout the distribution system, with the only exceptions being the unique provisions for retail outlets found in § 80.527 (e). Compliance is facility-based, so the fact that a company serves multiple roles does not matter.*

4.21: Define "distributor" and "end user" from a reporting and recordkeeping standpoint.

A: *A "distributor" is any entity in the distribution system -- they may or may not be named specifically as a pipeline, terminal, barge, rail system, etc. The activity that the distributor is engaged in will dictate appropriate reporting and recordkeeping requirements.*

An "end user" as addressed in the presentation, "Recordkeeping and Reporting for Nonroad Diesel Fuel", was specifically aimed at recordkeeping requirements for ultimate consumers in Alaska (§§ 80.600(f) & (g) and 80.554(a)(4) & (b)(5)).

4.22: EPA's 2004 refiners' pre-compliance report concludes that there will be an adequate supply of ULSD (at the refinery gates); however, the report does not address the distribution of ULSD. Will ULSD be available in all parts of the country?

> *A:* *Based on the information that refiners provided in the pre-compliance reports, as well as other discussions, we are confident that ULSD will be available in all areas of the country. Unfortunately, to respect confidential business information, we cannot provide more specific information in the summary and analysis of the pre-compliance reports on the exact locations that refiners would be producing ULSD. We are currently working on a proposal for a plan for Alaska that will address the unique situation found there.*

4.23: Will refiners produce No. 1 diesel fuel at 15 ppm sulfur? What other cold weather-gelling strategies are available to the end-user?

> *A:* *Yes, some refiners will produce No. 1 diesel fuel. There are also various other cold-flow improver additives currently on the market as well.*

4.24: Section 80.592 specifies the data which must be kept and maintained for all diesel batches but does not require submission to EPA unless requested. Section 80.604 requires annual submission of each batch of NRLM diesel or heating oil. However, the June 29, 2004 Preamble, section V(G)(5)(b) requires batch reports to be submitted, but does not specify between Highway or NRLM diesel. Please confirm that there are not any additional quarterly or annual reporting requirements for individual batch data of Highway diesel (15 or 500 ppm). In addition, does EPA plan on requesting this data?

> *A:* *There are no additional requirements, and EPA does not plan to request additional data at this time; the only requirements are volume and designation information plus identification information.*

4.25: If a refinery receives a previously designated distillate (PDD), what options does the refinery have for blending the PDD and how are any changes in fuel designation handled? In other words, is there a provision similar to the gasoline rule where a previously designated material may be debited from a compliance pool using a negative volume? Does a similar provision exist in the distillate rule? Alternatively, is the refinery to handle PDD as a terminal and be subject to the same downgrade rules as the terminal?

A: *There is no similar provision to the RFG/anti-dumping regulations that would treat previously designated distillate fuel as a negative batch when it is blended with blendstock at a terminal blending facility or other refinery. A facility that receives previously designated distillate must be registered as a downstream facility and make appropriate records and reports regarding designation and tracking of the previously designated distillate, including any downgrading or other change in designation of that fuel volume. In addition, if the refiner blends blendstock to the previously designated distillate, it must, in its capacity as a refiner, properly designate that additional volume and is responsible for all refiner requirements for that added volume. For example, if 100 gallons of high sulfur blendstock is added to 100 gallons of 15 ppm highway fuel, and the resulting 200 gallons of product is sold as 500 ppm highway fuel, in its capacity as a downstream facility, the facility would have to account for the downgrade of the 100 gallons of 15 ppm fuel to 500 ppm highway fuel. In its capacity as a refiner, it would be responsible for producing 100 gallons of 500 ppm fuel. If the example above is changed such that the product blended is 100 gallons of previously designated high sulfur fuel to 100 gallons of 15 ppm highway fuel, then there is no refining activity; there is just a change in designation and downstream downgrade of the 15 ppm fuel that is used for the blending to 500 ppm highway fuel.*

4.26: Do the Credit Trading Area (CTA) regulations as defined in section 80.531 apply only to importers?

    A: *No, CTAs apply to both importers and refiners.*

4.27: Section 80.532(d) implies that credits cannot be traded from one CTA to another- does this restriction apply only to imported fuel or can credits generated by a refinery located in PADD 1 (CTA 1) trade credits with a refinery in PADD 2 (CTA 2)?

    A: *This applies to both refiners and importers. Motor vehicle diesel credits can only be traded within the CTA where they were generated. However (per the technical amendments made in our May 1, 2006 action), early motor vehicle credits generated that were generated in CTAs 1-7 may be used in any of the CTAs 1 through 5.*

4.28: In earlier Q&As for gasoline, EPA designated that the Virgin Islands and Puerto Rico are in PADD 6; Guam, American Samoa and the Northern Mariana Islands are in PADD 7. Since CTAs are nearly identical to the corresponding PADDs why has EPA chosen CTAs 6 and 7 for Alaska and Hawaii, and will EPA reconsider these designations?

    A: *The gasoline and diesel programs are very different and their corresponding CTA definitions are therefore different (and were configured to address each program's specific needs). While these PADDs and CTAs were set for gasoline, we did not necessarily intend to have the diesel PADDs and CTAs to mirror those set for the gasoline program.*

However, we do realize that CTAs and PADDs for these areas were inadvertently left out of the highway and nonroad diesel rulemakings. The May 1, 2006 technical amendments to the regulations (see Section 16 for a link to this action) clarified and assigned PADDs and CTAs for these areas– the U.S. Virgin Islands and Puerto Rico are assigned to PADD VI and CTA 1. (As Guam, American Samoa and the Northern Mariana Islands are considered exempt under the diesel program, these areas are not being assigned to a specific CTA. Further, Alaska and Hawaii remain assigned to CTAs 6 and 7, respectively.)

4.29: What value are the non-highway baselines that are discussed in § 80.533 since they are not used in any credit generation equations or other compliances?

    A: *Section 80.533 explains that non-highway baselines are generated and used by small refiners to comply with any of the options available to small refiners under § 80.554; and other refiners for the purposes of early credit generation.*

4.30: The preamble states (page 39061), "[the] *rule allows terminal operators and others to switch the designation of 500 ppm sulfur NRLM diesel fuel to highway diesel fuel on a temporary but not a cumulative basis over time.*"
a) What "others" is EPA referring to?
b) What is temporary?
c) What if a party has not switched 500 ppm highway diesel back into the 500 ppm NRLM market by 2012 when there is no longer a market for 500 ppm NRLM (when NRLM must be at 15 ppm)?

    A: a)   *The word "others" in the sentence refers to pipeline operators.*
                  b)   *"Temporary" is within the compliance period.*
                  c)   *The party will be subject to quarterly compliance until 2010, and annual compliance thereafter. Any noncompliance will be detected at the end of each compliance period.*

4.31: Considering the likely contamination of large amounts of ULSD and the permissive downgrading provisions, will ULSD continue to be the dominant fuel at the pump in 2006?

    A: *Yes. Anti-downgrading allowances for each facility are 20%. However most facilities will try to limit downgrading. For even the likely worst case facility, downgrade through normal contamination is expected to be less than 10%, and typically only a few percent or less. In total, downgrade due to contamination is expected to average less than 5%. Consequently, it should not significantly impact the availability of 15 ppm fuel.*

4.32:   How will the end-user know whether they are refueling with fuel in excess of 15 ppm sulfur?

A: *The regulations require that all fuel pumps be labeled to inform the user which fuel they are dispensing into their vehicle. These labels will state which engines/vehicles that the specific fuel is suitable for use in. These labels will match comparable labels on 2007 and later highway vehicles. Distributors will likely have tested fuel batches at retail to ensure that the fuel meets the applicable standard that is on its pump label.*

4.33:   If a retailer or wholesale purchaser-consumer does not dispense early credit 15 ppm sulfur cap motor vehicle diesel fuel, is June 1, 2006 the effective date for the pump labeling standards at § 80.570 per § 80.500(e)?

A: *Pumps which dispense on-road diesel containing 15 ppm sulfur or less are required to be labeled as such starting June 1, 2006.*

4.34:   Under 40 CFR § 80.592(b) what is meant by a batch of diesel fuel and designating a batch number? 15 ppm sulfur content highway fuel will typically be produced continuously as the output from a distillate hydrotreater and is not blended like gasoline. The concept of batch numbering and batch volumes does not seem to make sense based on the production of this product. Please provide clarification.

A: *Under 40 CFR § 80.2(nn), a batch of motor vehicle diesel fuel "means a quantity of diesel fuel which is homogeneous with regard to those properties which are specified for motor vehicle diesel fuel under subpart I of this part"; a batch of motor vehicle diesel fuel is thus a discreet and identifiable quantity of such homogenous diesel fuel. Under the highway and nonroad diesel rules (at section 80.502(d)), production of a 'batch' has further been defined as the volume that is delivered to the next entity/facility. The batch exists at the moment of transfer, so your batch reports need to be tied to volumes delivered out of your facility. For refiners with continuous streams, volumes and properties should be measured periodically, and it would be best to apply the measurements to the volume produced most closely to the time of the measurement. A refiner that currently uses 'certification tanks' may perform testing (for designations) in these tanks, however the volumes that need to appear on batch reports/PTDs are the volumes that are delivered.*

        Assignment of batch numbers will be done the same way under the highway diesel sulfur rule as is done under the RFG and anti-dumping regulations. See §§ 80.592(b)(2) and 80.65(d)(3).

4.35:   Diesel fuel retailers will not be making 15 ppm motor vehicle diesel fuel; we will only be selling it to customers for use in their diesel-powered vehicles. What are a diesel retailer's primary regulatory responsibilities under the 15 ppm highway diesel fuel program?

A: *Responsibilities are similar to those under other fuels rules. While the responsibility for producing or importing compliant 15 ppm sulfur content highway diesel fuel resides with refiners and importers, retailers and all other parties in the distribution system share responsibility for assuring that diesel fuel subject to the 15 ppm standard is not contaminated or commingled with other products, such as distillates having a sulfur content greater than 15 ppm. At the retail dispenser, if the fuel subject to the 15 ppm sulfur standard does not meet the standard, it is in violation, subject to a 2 ppm test result adjustment applied to test results at downstream facilities. All parties in the distribution system are presumed liable for such violations, including the retailer (§ 80.612). Each party may establish a defense to the violation under § 80.613. A retailer may establish a defense by showing it did not cause the violation and that product transfer documents account for the fuel in violation and indicate the violating product was in compliance when it was under the retailer's control.*

In addition, retailers and wholesale purchaser-consumers are required under the regulations to post labels on each diesel fuel pump stand, notifying customers of the type of fuel dispensed by each dispenser. The required language for these labels is prescribed by § 80.570.

Per §80.610, retailers are prohibited from introducing, or permitting, the introduction of noncomplying fuel into motor vehicles. Per § 86.007-35, model year 2007 and later diesel powered vehicles, which will require fuel meeting the 15 ppm sulfur standard, will have a label on the filler inlet and on the dashboard stating "Use Ultra Low-Sulfur Diesel Fuel Only" or "Ultra Low-Sulfur Diesel Fuel Only." A retailer would clearly be liable for the action of a retail employee introducing 500 ppm diesel fuel into a vehicle equipped with such a label. Misfueling is addressed in more detail below.

Retailers must retain all product transfer documents, which identify the applicable standard of the fuel, and must assure that this fuel is delivered into the proper storage tank for sale to customers (see §§ 80.590, 80.591 and 80.592). If a misdelivery does occur, retailers must immediately cease sales from the dispensers served by the impacted storage tank and assure that the product in the tank is brought back into compliance before resuming sales.

Retailers should train their employees, and work closely with distributors and carriers who supply product to their outlets, so that the tank truck drivers will have needed information about what types of product are sold at each outlet and which storage tank corresponds to each type of product.

The truck distributors themselves should in turn work closely with the retail operations, and ensure drivers know which product should be delivered to each station, and what storage tank each type of product should be delivered to. Truckers should ensure that high sulfur products previously carried in truck compartments have been completely drained from the truck compartments and hoses before 15 ppm product is loaded into those compartments.

Retailers are also subject to downgrading limitations of the regulations (§ 80.527). This is addressed in more detail elsewhere. Retailers who downgrade 15 ppm highway fuel to 500 ppm highway fuel must keep records that demonstrate compliance with the limitations and requirements of the downgrading provisions (see § 80.527(g)).

4.36:   How do you indicate whether a batch of fuel is marked or unmarked?

A: *The product's accompanying PTD should state if the product is marked or not. Section 80.590(a) details the information that must be included on the product transfer document that accompanies any transfer in custody or title of MVNRLM diesel fuel or heating oil. Section 80.590(a)(6)(ii) states that the PTD must contain "an accurate and clear statement of the applicable designation and/or classification under section 80.598...and whether the fuel is dyed or undyed, and for heating oil, whether the fuel is marked or unmarked."*

Further, § 80.590(a)(7) contains additional requirements regarding the PTD language for the transfer of fuel batches of a given fuel designation/classification from one facility to another in the fuel distribution system where diesel fuel is taxed, dyed or marked. These requirements also apply for any subsequent transfers except for when the fuel is dispensed into an end-user's equipment.

4.37:   Can you clarify the PTD product code recommendation/requirements? Is there a specific product code standard that EPA would like us to use in reporting? Also, must the specific fuel designation be in the actual product code, or can it just be on the PTD somewhere?

A: *We will no longer be requiring the fuel sulfur level (15, 500, or >500) to be in the code of a PTD as long as the sulfur level is noted on the PTD somewhere. This was changed with our November 2005 technical amendment to the regulations (see Section 16 for links to all of the technical amendment actions). Please note that for transfers to truck carriers, retailers, or wholesale purchaser-consumers, product codes may not be used (§ 80.590 (d)).*

4.38:   My company currently sells distillate blendstocks. Can we continue to sell blendstocks given the new regulations?

A: *Yes, you may continue to sell blendstocks. In addition, it would also be in your best interest, for defense purposes, to identify any product that you ship.*

4.39: Has EPA developed a program that allows EPA to check balances between parties on hand-offs? If so, could you provide the same tool to industry to enable companies to run the tool to reduce errors before submission?

A: *No, EPA is not developing software to check hand-off balances. We will rely on database reports to identify potential hand-off reporting problems. Specifically, we will be looking at summaries of transactions, for a given quarter, between individual hand-off partners (shipper and receiver) to verify that the product volume shipped equals the product volume received, with zero tolerance.*

4.40: The Interstate Commerce Act requires shipper and volume confidentiality. Do the diesel fuel rules protect these requirements?

A: *While the EPA rules regarding confidential business information were not written for the Interstate Commerce Act, we do protect the confidentiality of submissions to EPA and do not give confidential company information out to others. If a company is concerned about the protection of its information, it may mark all reports and other submissions to EPA as "CBI" (Confidential Business Information).*

4.41: Biodiesel manufactured from palm oil has a red tint. Would this be problematic with respect to compliance with EPA red dye requirements (significant quantities of palm oil are available and the price is attractive)?

A: *EPA requires that fuel marketed as highway diesel fuel contain no visible evidence of dye solvent red 164. There is no regulatory requirement that would preclude that use of palm oil-derived biodiesel as a highway diesel fuel blendstock. EPA and other interested parties could use spectrographic analysis in the field to determine whether the source of a red tint to diesel fuel resulted from contamination with red dye 164 or from another source. Furthermore, we do not believe that this will be an issue, as biodiesel will not go through pipelines.*

4.42: All of the fuel that my terminal will be receiving is going to be 15 ppm fuel. The regulations at § 80.598(a)(3)(iv) state that "prior to June 1, 2009, all 1 ppm sulfur MVNRLM diesel fuel must be designated as motor vehicle diesel fuel." How do I provide NRLM fuel for my customers if I am only receiving 15 ppm fuel but cannot designate (and I presumably cannot inject dye, either) any of it as NRLM?

A: *Section 80.598(a) states the designation requirements for refiners and importers. Refiners and importers may not designate 15 ppm fuel that they produce as NRLM prior to June 1, 2009 (except for qualified small and GPA refiners that*

*will be producing all of their MVNRLM as 15 ppm under the "gas-for-diesel"
option). Facilities downstream of the refinery may redesignate 15 ppm highway
diesel fuel they receive to NRLM if they so choose.*

4.43:   Can EPA give some direction regarding, or an example of, an acceptable pump label?

   *A: While it is possible for labels to look somewhat different in regards to color
           scheme, maximum font size, etc., all labels must meet the following requirements
           (unless otherwise approved by EPA) specified in §§ 80.570-80.574:*
           - *text must be bold, block letters with a minimum font size of 24-point*
           - *the font color and background color must contrast*
           - *use the exact language stated in quotes*

           *Sample labels are available on the Clean Diesel Fuel Alliance (an alliance of
           government, industry, and environmental groups aimed at advancing education
           on the diesel programs) website at www.clean-diesel.org. Please note that these
           sample labels do have different font sizes than what is listed in the regulations,
           but they have been approved by EPA as acceptable alternate labels.*

4.44:   In § 80.530(a)(5), it states that the annual compliance period is June 1, 2006 through June
        30, 2007. However, § 80.599(a)(1) states that the annual compliance period is June 1,
        2006 through May 31, 2007. Which is correct?

   *A: Both are correct. The first annual compliance period for refinery production (for
           compliance with the Temporary Compliance Option, small refiner, and GPA
           refiner provisions) is June 1, 2006 through June 30, 2007. This date was chosen
           to coincide with refinery fiscal years. Therefore, § 80.530 (the section that deals
           with the TCO) and all sections regarding compliance for small and GPA refiners
           state June 30, 2007 as the end of the first compliance period. Section 80.599 lists
           the D&T quarterly and annual compliance periods. The first annual compliance
           period for D&T is May 31, 2007, which corresponds with the start date of the
           NRLM program on June 1, 2007.*

           *All parties in the D&T system must report their D&T compliance, as stated in §§
           80.601 and 80.604, using the compliance periods stated in § 80.599 (with the first
           compliance period being from June 1, 2006 through May 31, 2007). Refiners
           also need to report their TCO 80/20 compliance, per § 80.593, using the
           compliance periods stated in § 80.530 (with the first compliance period dates of
           June 1, 2006 through May 31, 2007).*

5.   Cetane Index/Aromatics Requirements

5.1:   In § 80.510(h), does EPA really mean that the cetane index (CI) doesn't have to be met if sulfur is met with credits?

A: *The cetane or aromatics requirement only applies to NRLM fuel that is required to meet either the 15 ppm or 500 ppm std. If fuel is produced to uncontrolled sulfur levels through, for example, credits or hardship provisions, that fuel does not have to be controlled to meet EPA's cetane or aromatics standards either. However, industry standards will still apply.*

5.2:   How can heating oil be used as NRLM when heating oil does not necessarily meet cetane and aromatics requirements?

A: *The rule allows distillate designated as heating oil to be redesignated as high sulfur NRLM provided that during any compliance period there was no net shift of heating oil into the NRLM market (per § 80.598(b)(9)(viii)). Please refer to the volume balance requirements for high sulfur NRLM and heating oil contained in 40 CFR 80.599(c). To meet these volume balance requirements, for any volume of heating oil redesignated as high sulfur NRLM during a compliance period, an equal volume of high sulfur NRLM would typically need to be redesignated as heating oil during that same compliance period. Any volume of heating oil redesignated as high sulfur NRLM would be required to meet the applicable cetane and aromatics requirements.*

5.3:   Cetane index does not work for alternative diesel formulations- diesel with 10% ethanol must be tested by Cetane number. Do you plan to make a technical correction?

A: *Both the highway diesel (at §80.520(a)(2)) and nonroad diesel (at §80.510(a)(2)) rules contain a requirement for cetane index <u>or</u> aromatic content standard. While cetane indexes do not work for alternative diesel formulations like "e-diesel", aromatic content can be used. As such, we do not believe that a technical correction would be needed.*

6. Sulfur Standards

6.1: In § 80.510(h), should it be "or" rather than "and". As currently written, only get exemption if credits, SR, hardship, and temporary relief?

A: Yes, it should say "or". This was corrected in a technical amendment to the regulations (which can be found on EPA's web site at: http://www.epa.gov/otaq/regs/fuels/diesel/diesel.htm#regs, July 7, 2005).

6.2: Have refiners based their production forecasts on 15 ppm ULSD, or some much lower pipeline spec?

A: Our understanding is that production forecasts are set based on a range of factors and as such, different refineries have different targets.

6.3: If 15 ppm kerosene may be blended with 15 ppm ULSD, does the kerosene also have to be certified as #1 ULSD?

A: No, although 15 ppm kerosene not previously certified as #1 ULSD will count against the party, provided that its highway volumes do not increase. Consequently, we expect that most #1 15 ppm will be designated as motor vehicle diesel fuel.

6.4: Should 15 ppm max sulfur No. 1 diesel or dual certified Jet/No. 1 diesel that is designated as 15 ppm diesel fuel when leaving the refinery be counted as 15 ppm diesel fuel in determining compliance with the 80% 15 ppm diesel fuel refinery production requirement for the Temporary Compliance Option?

A: Yes, if a refinery designates either No. 1 diesel or dual certified Jet/No. 1 diesel under as motor vehicle diesel fuel meeting the 15 ppm sulfur standard under § 80.520(a)(1), it is counted as 15 ppm diesel fuel in determining compliance with the Temporary Compliance Option.

6.5: Subsequent sale of the 15 ppm max sulfur dual certified kerosene as commercial jet fuel is permitted, is not subject to any downgrading limitation, and does not impact compliance with the 80% 15 ppm diesel fuel production requirement for the Temporary Compliance Option?

A: Correct, if a refinery designates kerosene as motor vehicle diesel fuel, then sells the kerosene as commercial jet fuel, then the reclassified volume of kerosene is not subject to any downgrading limitation, and does not impact compliance with the 80% 15 ppm diesel fuel production requirement for the Temporary Compliance Option.

6.6: Many refiners currently make a jet fuel meeting <500 ppm requirements. If refiners make a jet fuel meeting 15 ppm requirements, there may be compatibility (safety or performance) issues for jet engine manufacturers. Have these manufacturers been consulted in this rulemaking process? If so, what is their position on use of 15 ppm jet fuel?

    A: *The highway and nonroad diesel programs do not require that refiners produce 15 ppm sulfur jet fuel and do not set sulfur limits for jet fuel that is not designated as motor vehicle diesel fuel. This would be their decision. EPA believes compatibility issues raised by 15 ppm jet fuel would be addressed by the refiners and customers involved.*

7. Additives/Kerosene

7.1: Will there be an allowance for other additives in ULSD with a sulfur content greater than 15 ppm– like cetane enhancer?

   A: Additives with a sulfur content greater than 15 ppm may be added by terminals to diesel fuel that is subject to the 15 ppm sulfur standard, provided that the terminal ensures that the finished fuel blend remains compliant with the 15 ppm sulfur standard. To meet their affirmative defenses to presumptive liability, terminals that use above 15 ppm additives in ULSD typically must conduct a sulfur test on the finished fuel post-additization (see 40 CFR 80.613(a)(1(vi)). Conductivity additives and the red dye used to comply with Internal Revenue Service (IRS) requirements currently can not be manufactured with a sulfur content less than 15 ppm. We did not want the standard affirmative defense provisions (of § 80.613) to present a barrier to the use of conductivity additives and red dye. Therefore, EPA instituted alternative affirmative defense provisions regarding the use of these additives (through technical amendments to the regulations in May 2006, at § 80.614) that are based on volumetric accounting reconciliation and a sulfur test prior to additization. For other diesel fuel additive types, there are currently additives available that comply with the 15 ppm sulfur specification. Thus, alternative affirmative defense provisions are not needed for additives other than red dye and conductivity additives.

7.2. Subsequent use of No.1 Diesel or dual certified kerosene for winter blending in 15 ppm diesel fuel is not an issue, but is winter blending into 500 ppm motor vehicle diesel fuel subject to the 20% downgrading limitation by each custody or title holder on an annual basis?

   A: Per § 80.527, the anti-downgrading limitations, as modified in the nonroad rule, only apply to #2 15 ppm diesel fuel.

7.3: Some refiners, and some distributors, blend additives into diesel fuel to improve performance. Is EPA taking any steps to assure that these diesel fuel additives do not contain sulfur at levels in excess of 15 ppm?

   A: The highway diesel fuel provisions require diesel additive manufacturers to label their additives, both on the container (in the case of additives sold to consumers-- see § 80.591(d)(1) and (2)), and on the PTD (see § 80.591(b)(2)), with information as to the sulfur level in the additive. If the additive contains more than 15 ppm sulfur, then it is the blender's responsibility to assure that the use of the additive does not result in diesel that exceeds the 15 ppm standard (see § 80.521). Additives sold for consumer use in model year 2007 and later highway vehicles, and 2011 and later nonroad engines, may not contain more than 15 ppm sulfur (see § 80.591(d)(2)).

7.4: It is a common practice for diesel fuel to be blended with kerosene to improve performance characteristics (particularly during cold weather), or other substances, such as used oil to dispose of waste products. Can such practices be continued under the 15 ppm diesel fuel program?

> A: *Under § 80.522, used motor oil, or used motor oil blended into diesel fuel, may only be used in the fuel system of model 2007 or later diesel motor vehicles if the vehicle or engine manufacturer has received a Certificate of Conformity under 40 CFR Part 86, that certification is explicitly based on emissions data, and the motor oil is added at a rate consistent with the Certificate of Conformity. Under § 80.525, kerosene that a kerosene blender adds (or intends to add) to 15 ppm sulfur diesel must meet the 15 ppm sulfur content standard, as indicated by product transfer documents or test results.*

7.5: Kerosene is used for residential heaters, city bus fleets, and jet fuel among other uses. It is also used to winterize diesel fuel and home heating fuel. Today, kerosene comes in two grades, < 500 ppm sulfur and < .3% (3000 ppm) sulfur. It may or may not be dual purpose, i.e. suitable for jet fuel use. To minimize grades, refiners will likely produce only one grade of winterizing fuel, 15 ppm nominal maximum. Will 15 ppm kerosene that is used to winterize 500 ppm diesel fuel be considered a downgrade and count toward the 20%?

> A: *No, 15 ppm kerosene that is used for wintertime blending will not be considered downgrading.*

7.6: The anti-static additive sulfur contribution to ULSD will be resolved and tracked through VAR. All other non-complying additives will require analysis. We have been asked by some customers to acquire the anti-stat additive and blend it into packages they already purchase. The packages will be less than 15 ppm but of course addition of the additive can make them exceed the 15 ppm limit. Thus the question becomes, will EPA allow this combination package to be resolved via the VAR methods or will testing be required?

> A: *Multi-functional additive packages that contain a >15 ppm anti-static additive and/or red dye may use the VAR method. However, the other components in the additive package must be <15 ppm. If any component of a multi-functional additive package other than red dye or an anti-stat is >15 ppm, then the VAR method may not be used (§ 80.614). However, if the entire package is under 15 ppm, then the VAR method is not required.*

7.7: There are no restrictions on blending 15 ppm kerosene with ULSD. Is this statement modified depending on how the 15 ppm kerosene is designated?

70

A: *Yes. Unless the 15 ppm kerosene was designated as #1 MV diesel fuel, it would count against the blender in meeting their highway volume balance requirement under the D&T regulations. We expect that most refiners will choose to designate 15 ppm kerosene as No. 1 MV diesel fuel.*

7.8: Kerosene with a sulfur level in excess of 15 ppm can be blended with ULSD up to a max of 1% volume provided it does not alter the quality specs of the ULSD (and sulfur level of the ULSD does not exceed 15 ppm). I assume because of the 1% volume limitation that it doesn't matter what the designation of the plus-15 ppm sulfur kerosene is. True?

A: *No, this is not correct. Kerosene that is blended with ULSD must meet the 15 ppm standard, as stated in § 80.525.*

Kerosene is not an additive ((per § 80.2(xx)), nor is it considered as such with respect to the 1% additive blending provisions. § 80.525 states that kerosene blended with MVNRLM must meet the applicable standard of the MVNRLM. The accompanying PTD must also state that the kerosene meets the applicable standard, and the kerosene blender must have test results indicating that the kerosene complies with that standard. (See also QA 7.4.)

7.9: There are no restrictions on blending 500 ppm kerosene into LSD. Is this statement modified depending on how the 500 ppm kerosene is designated?

A: *The only restriction is on a party's volume balance- the total volume of highway diesel fuel cannot go up, so blending 500 ppm kerosene will count against a party's volume balance requirement under the D&T regulations.*

7.10: Are there any rules regarding the blending of plus-500 ppm kerosene into LSD similar to the 1% volume rule for ULSD?

A: *As stated in 7.8, above, § 80.525 restricts the blending of plus-500 ppm kerosene into LSD. Further, the 1% volume rule does not apply, as kerosene is not an additive. Blending kerosene that exceeds the applicable fuel's standard would result in the need to redesignate the finished fuel. Such blending could also impact the blending party's highway diesel volume balance.*

7.11: 40 CFR 80.614 states that the alternative affirmative defense provisions are applicable for a static dissipater additive that has a sulfur content greater than 15 ppm but whose contribution to the sulfur content of the MVNRLM diesel fuel is less than 0.4 ppm at its maximum recommended concentration. The way I interpret this is that if your static dissipater additive contributes less than 0.4 ppm sulfur you can use the calculation method to prove compliance. If the static dissipater additive would contribute more than 0.4 ppm sulfur, this additive could still be used but compliance would have to be proven through testing as outlined in 40 CFR 80.613(a)(1)(vi). Is this correct?

A: *Yes, this is correct– static dissipater additives that contribute more than 0.4 ppm to the sulfur level of the finished fuel could still be used. However, one would need to follow the provisions of § 80.613 to establish affirmative defenses to presumptive liability. The alternate provisions of § 80.614 only apply to static dissipater additives that contribute less than 0.4 ppm to the sulfur level of the finished fuel and red dye that contributes less than 0.04 ppm to the sulfur level of the finished fuel.*

7.12: The regulations at 40 CFR 80.570 establish pump labeling requirements, however it is unclear whether or not these labels apply to pumps dispensing kerosene. The IRS regulations require labels for kerosene pumps.

Pumps that dispense dyed kerosene are labeled, pursuant to IRS regulations at 26 CFR 48.4082-2(a), with "DYED KEROSENE, NONTAXABLE USE ONLY, PENALTY FOR TAXABLE USE." Pumps that dispense clear kerosene that is untaxed must be blocked and labeled, pursuant to 26 CFR 48.6427-10(b)(1)(iii), with "UNDYED UNTAXED KEROSENE, NONTAXABLE USE ONLY." Are these labels adequate for EPA's pump labeling requirements?

A: *Kerosene meets the definition of diesel fuel since it is a distillate fuel that is suitable for use in diesel engines and it is considered No. 1 diesel fuel (40 CFR 80.2(x)). Since it is diesel fuel, the applicable sulfur standards and pump labeling requirements depend on how it is distributed and used.*

There are two situations presented. In one case the kerosene is dyed and distributed from pumps labeled "DYED KEROSENE, NONTAXABLE USE ONLY, PENALTY FOR TAXABLE USE" pursuant to the IRS labeling requirements of 26 CFR 48.4082-2. In the other case, the kerosene is undyed and distributed from a blocked pump that is labeled "UNDYED UNTAXED KEROSENE, NONTAXABLE USE ONLY" pursuant to the IRS labeling requirements of 26 CFR 48.4082-2.

In both cases, EPA will assume it is not intended for use or made available for use as motor vehicle diesel fuel absent evidence that shows otherwise. This kerosene would therefore not be subject to the sulfur standards for highway diesel fuel and would not be subject to the pump labeling requirements for highway diesel fuel at § 80.570(a) or (b). However if the kerosene was in fact sold, dispensed, or offered for sale to highway vehicles it would then be motor vehicle diesel fuel and the kerosene would be subject to the sulfur standards and the pump would be subject to the labeling requirements for highway diesel fuel.

Given the statement on the kerosene pumps that the diesel fuel is for "nontaxable use only," EPA will assume that this kerosene is being sold, dispensed, or offered for dispensing for nonroad equipment. The pump is therefore subject to the labeling requirements of § 80.570(c) unless the kerosene meets the sulfur

72

standards for highway fuel. While EPA would evaluate the specific circumstances of a situation to determine the applicability of the pump labeling requirements, it is reasonable to assume that these kerosene pumps are subject to the labeling requirements of § 80.570(c) absent clear evidence otherwise.

Under § 80.570(c), the pump labels need to contain certain specified information, such as the statement that the diesel fuel "may exceed 500 ppm sulfur." The pump label statements required by IRS do not satisfy the various requirements of § 80.570(c).

EPA does have authority to approve alternative pump labels, and has approved alternative labels for kerosene pumps. EPA would approve the following labels for kerosene pumps as an alternative to the labels specified in § 80.570.

(1) Alternate labeling language for 40 CFR 80.570(a):

For clear, taxed kerosene-

ULTRA-LOW SULFUR KEROSENE (15-ppm Sulfur Maximum)
Required for use in all model year 2007 and later highway diesel vehicles and engines.
Recommended for use in all diesel vehicles and engines.

For clear, untaxed kerosene at blocked pumps-

ULTRA-LOW SULFUR KEROSENE
(15-ppm Sulfur Maximum)
Undyed and untaxed.
Not for use in highway diesel vehicles and engines.

---

(2) Alternate labeling language for 40 CFR 80.570(b):

For clear, taxed kerosene-

LOW SULFUR KEROSENE (500-ppm Sulfur Maximum)
WARNING
Federal law *prohibits* use in model year 2007 and later highway vehicles and engines.
Its use may damage these vehicles and engines.

For clear, untaxed kerosene at blocked pumps-

LOW SULFUR KEROSENE (500-ppm Sulfur Maximum)
Undyed and untaxed.
Not for use in highway diesel vehicles and engines.
May Damage 2007 or later model year diesel engines.

73

(3) Alternate labeling language for 40 CFR 80.570(c):

NON-HIGHWAY KEROSENE (May Exceed 500-ppm Sulfur)
WARNING
Federal law *prohibits* use in highway vehicles or engines. Its use may damage these vehicles and engines.

---

8. Credits

8.1: § 80.599(b)(6) states "*Calculations in paragraphs (b)(4) and (b)(5) of this section may be combined for all facilities wholly owned by an entity.*" May this combination include facilities in different Credit Trading Areas? [§ 80.599(b)(6)]

> A: *Yes. The Credit Trading Area restriction on highway diesel fuel credit generation and use only applies to refinery production and not to subsequent movement of the fuel through the distribution system or downstream compliance under the designate and track provisions.*

8.2: In § 80.535, the regulatory language covers generation of high sulfur credits for early production of 500 ppm NRLM and generation of 500 ppm credits for early (2009) production of 15 ppm NRLM. Can a refiner with some facilities that will be producing 15 ppm NRLM starting in 2006 generate high sulfur credits for early production of 15 ppm NRLM in 2006, as long as it does not double count these credits?

> A: *No- in order to receive NRLM credit, the fuel must be designated as NRLM. Further, to receive high sulfur credits, the fuel must at least meet the 500 ppm sulfur standard. Per §80.598 (a)(3)(iv), any fuel designated as 15 ppm until June 2009 must be designated as motor vehicle diesel fuel in order to maintain the intended benefits and integrity of the highway program. Such fuel can always be used for NRLM purposes. However, in order to receive NRLM credit in the situation described in the question, the fuel would have to be designated as 500 ppm NRLM by the refiner even though it met the 15 ppm standard.*

8.3: For each of the four compliance periods between 6/1/2006 and 12/31/2009, an interpretation of § 80.530(a)(3)(i)(A) is the following:
$V_{500} \leq [0.2 \times (V_{15} + V_{500})]$ + credits from the same Credit Trading Area, where the units are gallons.

Is this algebraic interpretation correct?

> A: *Yes. For each compliance period between 6/1/2006 and 12/31/2009, inclusive, the volume of diesel produced or imported at 500 ppm is limited to the sum of any credits properly generated and used by the refiner or importer, or properly transferred to and used by the refiner or importer from within the same Credit Trading Area, plus 20 percent of the total volume of V15 and V500 diesel produced or imported. There are also certain additional requirements on the generation and use of credits. See §§ 80.531 and 80.532.*

8.4: For the compliance period from 1/1/2010 through 5/31/2010, an interpretation of § 80.530(a)(3)(i)(B) is the following:

$V_{500} \leq$ credits from the same Credit Trading Area, where the units are gallons
Is this interpretation correct?

A: Yes. *For the period 1/1/2010 through 5/31/2010, the volume of diesel produced or imported at 500 ppm is limited to the sum of any credits properly generated and used by the refiner or importer or properly transferred to and used by the refiner or importer, from within the same Credit Trading Area. There are also certain additional requirements on the generation and use of credits. See §§ 80.531 and 80.532.*

8.5:   Carryover of a small deficit is permitted, if two conditions are met in the immediately following compliance period. There is a limit on the size of this deficit. § 80.530(a)(6): ". . . However, for any compliance period prior to and including 2009, a refiner and importer may exceed the volume limit in paragraph (a)(3) of this section by no more than 5 percent of the volume of V_t of diesel fuel produced or imported during the compliance period, . . ." An interpretation of this 5 percent limit is the following:
$$V_{500} > [0.2 \times (V_{15} + V_{500})] + \text{credits},$$
if two conditions are met in the immediately following compliance period and if $V_{500} -$ ($[0.2 \times (V_{15} + V_{500})] + \text{credits}) \leq 0.05 \times (V_{15} + V_{500})$, where the units are gallons and the credits must be generated in the same Credit Trading Area that they are used.
Is this algebraic interpretation correct?

A: *This algebraic formula correctly calculates the volume of deficit that can be carried over under § 80.530(a)(6). The volume of 500 ppm diesel produced or imported in any compliance period may exceed the volume limit of 500 ppm diesel allowed in that compliance period provided that 1) the volume of 500 ppm diesel does not exceed 5 percent of the total volume of V15 and V500 diesel produced or imported during the compliance period, and 2) in the succeeding compliance period, the refiner or importer meets the volume limit for that year on V500, and produces or imports 15 ppm diesel, or uses credits, equal to the volume of the exceedance in the preceding compliance period.*

8.6:   The definition of V_{500} is included in § 80.531(a)(2): "V_{500} = the total volume in gallons of diesel fuel produced or imported that is designated under § 80.598(a) as motor vehicle diesel fuel and subject to the 500 ppm sulfur standard under § 80.520(c) plus <u>the total volume of any other diesel fuel (not including V15, diesel fuel that is dyed in accordance with § 80.520(b) at the refinery or import facility where the diesel fuel is produced or imported, or diesel fuel that is designated as NRLM under § 80.598(a)) represented as having a sulfur content less than or equal to 500 ppm</u>." (emphasis added) Please explain the underlined text above. What is this "other diesel fuel"?

A: *"Other diesel fuel" means any distillate products that meet the definition of diesel fuel, such as kerosene, that is represented to have a sulfur content less than or equal to 500 ppm. Any volume of refined distillate products containing less than*

500 ppm sulfur that is not designated as motor vehicle diesel fuel, should not be included in the total volume of motor vehicle diesel fuel.

8.7: The federal temporary compliance option (TCO) includes a credit trading program with a relevant restriction: if a refinery produces diesel fuel for a <u>state</u> 15 ppm sulfur cap program for a greater volume than the federal requirement, then that fuel is <u>excluded</u> from the federal credit program. See § 80.531(a)(5)(iv). This restriction applies to both motor vehicle diesel fuel produced in that state or imported directly into that state.

8.7a: The Texas Low Emissions Diesel (TX LED) rule (with its 15 ppm sulfur cap for highway and non-road diesel in three ozone nonattainment areas and 95 eastern/central counties beginning on June 1, 2006) possibly excludes a significant volume of ULSD from the federal TCO credit generation program beginning on June 1, 2006. On October 15, 2001, EPA Region 6 approved the TX LED program as necessary for Houston-Galveston ozone attainment; this approval was published in the *Federal Register* on November 14, 2001 (66 FR 57196). If it meets the requirements in § 80.531, will 15 ppm TX LED be included in the federal TCO credit program beginning on June 1, 2006 because, although the 15 ppm TX LED rule includes non-road diesel fuel and excludes the federal temporary compliance option, the 15 ppm TX LED volume is not expected to be greater than the federal 15 ppm diesel fuel requirement in the state? Will early 15 ppm sulfur cap TX LED produced before June 1, 2006 qualify as an early federal TCO credit if it meets the requirements in § 80.531? Or will § 80.531(a)(5)(iv) exclude all motor vehicle diesel (TX LED and other diesel) produced in or imported into Texas beginning on June 1, 2006 from the federal TCO credit program because "no motor vehicle diesel fuel produced in that state or imported directly into that state may generate credits under this subpart, ..." even though this Texas diesel sulfur regulation will only apply to part of the state?

A: *The highway diesel program limits the ability to generate credits for fuel produced or imported directly into a State with its own diesel fuel requirement that requires a greater volume of 15 ppm or lower fuel than the Federal program. Upon evaluating the Texas LED program in the context of § 80.531(a)(5)(iv), we have concluded that, while the Texas program is more stringent for a portion of Texas, it does not appear to require a greater volume of fuel on a statewide basis to be 15 ppm than will be required under the Federal program. As a result, in this specific instance, the limit on credit generation in § 80.531(a)(5)(iv) does not apply to the Texas program.*

8.7b: On February 24, 2000, CARB approved the Public Transit Bus Fleet Rule and Emissions Standards for New Urban Buses. This includes a diesel fuel sulfur cap of 15 ppm for transit agencies effective July 1, 2002. Does this CARB 15 ppm diesel fuel for public transit bus fleets qualify for the federal TCO credit program, if the conditions in § 80.531 are met, because California is not subject to the waiver of federal preemption provisions in the Clean Air Act? Or does § 80.531(a)(5)(iv) exclude all motor vehicle diesel fuel produced in California from the federal TCO credit program because "no motor vehicle diesel fuel produced in that state or imported directly into that state may generate credits under this subpart, . . ." even though this CARB diesel sulfur regulation only applies to public transit bus fleets?

> A: *The § 80.531(a)(5)(iv) restriction does not apply to fuel produced prior to June 1, 2006. Credit for fuel produced prior to then is subject to the provisions in § 80.531(b) and (c). Given that this program covers less than half of the diesel fuel sold in California, it would also not be considered to require a greater volume of 15 ppm diesel fuel than the Federal program under the provisions of § 80.531(a)(5)(iv).*

8.7c: On September 15, 2000, the South Coast Air Quality Management District (SCAQMD) adopted a diesel sulfur content cap of 15 ppm effective on October 1, 2005. Will this SCAQMD 15 ppm diesel fuel qualify for the federal TCO credit program, if the conditions in § 80.531 are met, because California is not subject to the waiver of federal preemption provisions in the Clean Air Act? Or will § 80.531(a)(5)(iv) exclude all motor vehicle diesel fuel produced in California beginning on October 1, 2005 from the federal TCO credit program because "no motor vehicle diesel fuel produced in that state or imported directly into that state may generate credits under this subpart, . . ." even though this SCAQMD diesel sulfur regulation will only apply to part of the state?

> A: *Please refer to responses a and b, above. The SCAQMD issue is similar to the Texas LED issue. Given that this program covers less than half of the diesel fuel sold in California, it would also not be considered to require a greater volume of 15 ppm diesel fuel than the Federal program under the provisions of § 80.531(a)(5)(iv). However, since CARB has subsequently adopted a similar state-wide program for 15 ppm diesel fuel (for highway and off-highway use), it requires a greater volume of 15ppm diesel fuel than the Federal program and the restrictions in § 80.531(a)(5)(iv) apply.*

8.8:   Credits must be designated by refinery or importer-port of import, year of generation, and Credit Trading Area (1-7) of generation. In 2005 and 2006, is it necessary to further designate credits generated before June 1 versus after May 31 because § 80.531(c) has an early credit period from June 1, 2005 through May 31, 2006?

> A: *For reporting purposes, it is not necessary to distinguish between credits generated before June 1, 2006 (early credits) versus credits generated after June*

1, 2006, unless the early credits are going to be used in a different CTA from which they were generated in (per the changes made in our May 2006 technical amendments to the regulations). Specifically, § 80.593 states that annual compliance reports must contain information regarding credits for each refinery or, in the case of importers, by Credit Trading Area. For record keeping purposes, refiners and importers must keep information on credits separately for each calendar year compliance period and separately for each refinery (or in the case of importers, for each Credit Trading Area). However, § 80.592(b) also specifies that refiners and importers must keep documentation on the calculations used to determine the number of credits generated. Since there are three ways to generate credits under the program, credit calculations must be appropriate for the given time period in which the credits are generated.[1]

8.9: For reporting the percentage of motor vehicle diesel fuel produced meeting the 15 ppm sulfur standard after the inclusion of any credits, an interpretation of § 80.593(a)(4) is the following:

$$\% = 100 \times [(V_{15} + \text{credits})/(V_{15} + V_{500})]$$

where the credits must be generated in the same Credit Trading Area that they are used. Is this algebraic interpretation correct?

A: *Yes, this algebraic formula correctly calculates the percentage of fuel that must be reported under § 80.593(a)(4), where V15 and V500 have the meanings provided in § 80.531(a)(2), and "credits" means credits that are properly generated and used under the regulations.*

8.10: Adequate volume of 15 ppm highway diesel fuel produced today for the testing of the distribution system is key to resolving uncertainty in a timely manner and working out implementation issues. What is EPA doing to encourage refiners to produce test batches of 15 ppm highway diesel fuel so that downstream parties can test models and assumptions while there is still time to modify systems prior to 2006? Is it possible to credit current production?

A: *To ensure a smooth transition to large-scale production and distribution of 15 ppm highway diesel fuel beginning June 1, 2006, the program allows for refiners to generate early credits. From June 1, 2001 through May 31, 2005 refineries*

---

[1] 1. From June 1, 2001 through May 31, 2005 refineries and importers may generate credits based on the volume of 15 ppm sulfur highway diesel fuel that is used in vehicles with engines that are certified to meet the model year 2007 heavy-duty engine PM standard under 40 CFR 86.007-11.
2. From June 1, 2005 through May 31, 2006 refineries and importers may generate credits based on the volume of 15 ppm sulfur highway diesel fuel that is dispensed at retail outlets or wholesale purchaser-consumer facilities.
3. From June 1, 2006 through December 31, 2009 refineries and importers may generate credits based on the volume of 15 ppm sulfur highway diesel fuel produced above the 80 percent threshold.

and importers may generate credits based on the volume of 15 ppm sulfur highway diesel fuel that is used in vehicles with engines that are certified to meet the model year 2007 heavy-duty engine PM standard under 40 CFR 86.007-11. From June 1, 2005 through May 31, 2006 refineries and importers may generate credits based on the volume of 15 ppm sulfur diesel fuel that the produce. Further, we are allowing early credits that were generated in CTAs 1-7to be used in CTAs 1-5.

Under the existing regulations, diesel fuel used to test the distribution system does not generate early credits unless it meets the above criteria. However, based on recent discussions with interested stakeholders, the Agency is also investigating whether it would be appropriate to allow early credits for test batches of 15 ppm sulfur diesel fuel which, while starting out as 15 ppm may end up at a higher sulfur level prior to sale to the end user. The Agency is in the process of developing a proposal to allow this.

8.11: Can credits generated by meeting the 15 ppm highway diesel sulfur standard early be used in the gasoline program?

A: *No. However, the regulations at § 80.540 allow a GPA refiner to extend its sulfur standards for gasoline from December 31, 2006 to December 31, 2008 if the GPA refiner produces 95% of its on-road diesel as 15 ppm diesel (at a volume that is at least 85% of its baseline volume) by June 1, 2006. Also, the regulations at § 80.553 allow a small refiner to extend its sulfur standards for gasoline to December 31, 2010 if the small refiner produces 95% of its on-road diesel as 15 ppm diesel by June 1, 2006 (at a volume that is at least 85% of its baseline volume).*

9. Product Transfer Documents (PTDs)/Bills of Lading (BOLs)

9.1: Please verify that the "product document requirements" of §80.590 do not apply to transfers where an entity maintains custody of a batch of diesel fuel from one place in the distribution system to another place (e.g., from a refinery to a pipeline or a pipeline to a terminal), all owned by the same entity. It is understood that records must be retained under §80.600(a)(7).

   A: Product transfer documents must be exchanged when a product changes custody. A facility is defined as a location or series of location where custody does not change, therefore product transfer documents do not need to be exchanged between locations within a single facility.

9.2: Must BOL's contain the exact words of this section? (§§ 80.590, 80.598)

   A: For parties upstream of the retail outlet or wholesale purchaser-consumer, the PTDs required under the regulation may use product codes, except that the sulfur content standard must appear in numeric form (15, 500 or >500). For PTDs to retail outlets or wholesale purchaser-consumers, the exact language must be used (unless alternative language has been approved by EPA).

9.3: Can codes be used to designate facility IDs on PTDs, or does the entire facility ID need to be on the PTD?

   A: To ensure accurate handoff accounting under D&T, PTDs require the disclosure of the full 9 character Facility ID (EPA assigned 4 character Company ID plus EPA assigned 5 character Facility/Activity ID).

9.4: Are separate PTD documents required for diesel or will general procedures currently in place to meet gasoline PTD requirements be sufficient to meet PTD requirements for all grades of diesel?

   A: While there is substantial overlap in PTD requirements for gasoline and diesel, the requirements are not identical.

9.5: What data points will EPA want/need to allow pipelines to develop necessary PTD language?

   A: *The information required on commercial PTDs includes identification of the transferee and transferor (including names and addresses of the parties, and registration number of transferor and transferee (as amended in the July 7, 2005 technical amendment to the regulations), the volume of product, the date and location of the transfer, and identification of diesel fuel distributed by use designation (for use in motor vehicles, NR equipment, LM equipment, or NRLM equipment) and the sulfur standard to which the fuel is subject. PTDs must indicate the type of fuel- whether diesel fuel, heating oil, kerosene, exempt fuel, or other. PTDs must also state if fuel is No. 1 or No. 2; dyed or undyed fuel; and marked heating oil, marked LM, or unmarked fuel. Where a party delivers or receives fuel that has two different designations (but a uniform sulfur content), parties must use separate PTDs for each usage designation. At the point where fuel is taxed or dyed/marked, and for subsequent transfers, the PTD must indicate the applicable fuel uses as well as the standard. This is not an exhaustive list of the requirements, a complete and detailed description of the specific requirements is located in § 80.590.*

9.6:   On the designate and track approach, is any specific information required to be written on the Bill of Lading (BOL) at the truckload rack?

   A: Section 80.590(d) provides that product codes may be used if such codes are clearly understood by each transferee (so long as the sulfur content standard, stated numerically in parts per million, is included), except that the information required by § 80.590 to be conveyed to truck carriers, retailers, wholesale purchaser-consumers and mobile refuelers must be stated verbatim. See also § 80.590(g).

9.7:   If a retailer or wholesale purchaser-consumer does not dispense early credit 15 ppm sulfur cap motor vehicle diesel fuel, is June 1, 2006 the effective date for the product transfer document (PTD) requirements at § 80.590 per § 80.500(e) or § 80.592(a)(1) or § 80.530(a)(2)?

   A: The PTD requirements apply to all motor vehicle diesel fuel under the highway diesel rule (both 15 ppm and 500 ppm) as of June 1, 2006. Prior to June 1, 2006, the PTD requirements apply only to transfers involving early credit fuel. Under § 80.590(h) and (i), starting June 1, 2001, any highway diesel fuel transfers that are subject to the early credits provisions of § 80.531(b) or (c) are subject to the PTD requirements. The 5 year retention period under § 80.592(a) applies to all required PTDs under the rule, including PTDs required under the additive provisions (see § 80.591).

9.8:   Some pipelines operate a fungible batch system and some operate a segregated batch system. A shipper on a segregated batch system receives the batch that it tendered at the origin point. The pipeline has the responsibility to maintain batch quality while the batch is in its custody. If the pipeline delivers to third party tankage, the terminal operator is responsible for segregating interface and delivering to tankage. It is unclear where the pipeline's responsibility begins and where it ends, which could drastically affect the pipeline's sampling and testing program. (How is responsibility divided under the current 500 ppm program?)

A: *Pipelines are carriers under the regulations and as such have responsibility to ensure the product meets the standards while the pipeline has custody or title, and to provide appropriate product transfer documents to any downstream parties to whom it transfers custody. The responsibility to account for downgrade is a function of the point at which custody of the fuel is deemed to take place. This point is a function of contracts between the different entities. Each entity is responsible for all downgrading that occurs when it has custody.*

9.9:   At truck loading terminals what additional information needs to be provided to implement ULSD? Specifically, what information needs to be printed on the tickets and what information needs to be included in the quarterly report?

A: *The information detailed in §§ 80.590 and 80.591 needs to be included on all PTDs. The information that must be included in quarterly reports is stated in detail in § 80.601(a). PTDs can serve as commercial documents (such as "tickets"), as long as they contain all of the information that is required by EPA.*

10.   Documentation, Reporting, and Recordkeeping

10.1:   Section 80.600(a)(10) says, "*Any refiner or importer shall maintain copies of all product transfer documents required under §80.590. If all information required in paragraph (a)(6) of this section is on the product transfer document for a batch, then the provisions of this paragraph (a)(10) shall satisfy the requirements of paragraph (a)(6) of this section for that batch.*" Does that mean that if a refiner or importer has a PTD containing the information no other record is required? That seems unneeded. Does that paragraph mean something else?

A: *Section 80.600(a)(10) first requires that product transfer documents under § 80.590 be maintained by refiners and importers. It then states that if such product transfer documents contain all the information required under paragraph (a)(6) of § 80.600 for each batch, then no other separate recordkeeping is required for that batch in order to satisfy the requirements of paragraph (a)(6).*

10.2:   Must records be retained if only #2D 15 PPM is received? In general, are requirements of § 80.600(b) applicable if the distributor only receives #2D 15 PPM? [§§ 80.600(b)(1)(ii) and 80.600(b)(2)]

A: *Yes, records must be retained for this fuel (unless the fuel is dyed, marked, and/or taxed or in the Northeast/Mid-Atlantic Area or Alaska) for at least five years. In this case, the D&T reports would confirm that the distributor did not handle any other fuel and did not violate any of the restrictions on redesignating fuel. The D&T reports from this distributor would also be used in evaluating compliance by the parties before and after this party in the distribution system. This distributor is only one step in a chain of fuel hand-offs, and the D&T system is designed to use information from each party to a hand-off to ensure compliance by all of the parties in the chain of distribution. Further, parties should retain records for defense purposes.*

10.3:   a)   Are diesel batch numbers kept separate from gasoline batch numbers?
    b)   Must numbers be chronological and contiguous?
    c)   Are zero volumes allowed? [§ 80.602(b)(2)]

A: a)   *Yes. §§ 80.592, 80.593, 80.602(b)(2) and 80.604 require that the batch numbering conventions set forth in § 80.65(d)(3) be used for diesel fuel. The regulations require batch numbering for diesel fuel to be separate from batch numbering for purposes of the gasoline rules (RFG/anti-dumping, gasoline sulfur), using the same batch numbering conventions for diesel as are used for gasoline.*
    b)   *Numbers must be chronological and contiguous- under § 80.65(d)(3), the 1st batch for the year is number 1, with ". . . each subsequent batch during*

84

*the calendar year being assigned the next sequential number." Each
batch number also includes the EPA company and facility registration
number and the last 2 digits of the year in which the batch was produced.*

    c)   *Yes, zero volumes are allowed. For example this would be appropriate
where a batch is tested, then re-blended and retested before any of it is
transferred. Back-up records should be kept so that an adequate
explanation exists regarding why there was a zero volume.*

10.4:   A tank truck common carrier picks up a load of ULSD at a Shell terminal and delivers it to a Flying J truckstop. The carrier only provides a transportation service, does that carrier have any registration or reporting duties?

    A: *No, the carrier does not have registration or reporting duties as long as taxes
were assessed (in the case of highway fuel) or dye was added (in the case of
NRLM) at the terminal. There are recordkeeping and PTD requirements,
however. These requirements are listed in detail in §§ 80.590 and 80.592 of the
regulations.*

10.5:   A trucking company stores ULSD at its terminal for use in its own truck fleet. Does it have any registration or reporting requirements?

    A: *No, but there are recordkeeping requirements; and such an arrangement may be
considered, by definition, a "Wholesale Purchaser-Consumer".*

10.6:   Is it correct that below the rack, the primary recordkeeping mandate is production and retention of PTDs with no reporting obligations?

    A: *Yes, if requisite dye/marker has been added and/or taxes were paid (see QA 10.4,
above).*

10.7:   Do product codes have to contain the sulfur amount if a description accompanies the code? Will EPA provide examples of code structure?

    A: *The numeric designation of 15, 500, or >500 must appear on all product transfer
documents for MVNRLM. The rule (at § 80.590(d)) currently requires that the
numeric designations appear in the PTD's code, however, we intend to revise that
provision in a subsequent technical amendment to allow the designation to be
stated anywhere on the PTD, not just in the code (but the numeric designation
must be clearly marked on the PTD somewhere). We will also allow product
codes on PTDs to contain another 'symbol' in place of the numeric designation if
that symbol is defined somewhere on the PTD (e.g., to denote that a fuel meets the
15 ppm standard, "X" can be included in the code in place of "15", as long as it
is clearly noted on the PTD somewhere that "X = 15 ppm").*

10.8: What compliance and reporting requirements are there if a company throughputs highway diesel at a 3rd party's terminal?

      A: Under the D&T regulations, the custody holder is the responsible party, and must comply with all recordkeeping and reporting requirements. In this example, it would in most cases be the terminal owner. However, in order to ensure compliance for the terminal as a whole, they may need to place their own compliance requirement(s) on the lessee.

10.9: Must individual batch reports be submitted for highway diesel (15 or 500 ppm)? Is this requirement for NRLM diesel only?

      A: With respect to highway diesel, batch reports are required for participants in early credit generation and/or refiners covered under a Temporary Compliance Option, per § 80.581.

10.10: Is registration for an EPA number at a terminal required, and is it in addition to present TCN #?

      A: Yes, registration is required. The TCN # is used for the IRS system; the EPA D&T system is not based on the IRS model, but rather on our existing gasoline reporting programs. Facility ID numbers are assigned by facility, however, an aggregated facility or a single facility that serves many roles would need to state those activities (such as: refinery, importer, breakout (passthrough) terminal and/or pipeline, and truck loading terminal) in its reporting documentation.

      Please see the "Diesel Fuel Reporting Forms" page for registration forms and information, at:
      http://www.epa.gov/otaq/regs/fuels/dieselfms.htm

10.11: a) What is the difference in reporting requirements for additives containing greater than 15 ppm, as well as those containing less than 15 ppm, sulfur?
      b) Will batch recertification change based on additive sulfur content?

      A: a)   There is no reporting requirement for additives, only MVNRLM.
                  b)   Redesignation of a batch due to downgrading is limited to 20% of volume by facility. Additives having a sulfur content greater than 15 ppm that are used in ULSD must meet the standards and identification requirements of § 80.521(b).

10.12: What registration and reporting requirements exist for 3rd party terminals and other service providers that never actually own the product with regard to ULSD?

A: Under the designate and track regulations, custody holders are the responsible party for registration, recordkeeping, and reporting.

10.13: Since all entities in the distribution system will be required to report data, this creates double reporting as both the transferor and transferee would be reporting the same volume. Are there any plans or safeguards to be in place to allow a reporting entity to check and/or verify volumes that other entities have assigned to their entity or facility?

A: There are no plans/requirements being put in place by EPA. Businesses may handle such a program themselves, as they see fit, as part of a quality assurance program.

10.14: Has EPA defined the reporting application or process for the designated volumes reports? Will alternate methods, such as EDI or spreadsheet be allowed for submissions?

A: Please see the "Diesel Fuel Reporting Forms" page for applications, and all other reporting forms, at:
http://www.epa.gov/otaq/regs/fuels/dieselfms.htm
Alternate methods- spreadsheets or flat files- may be used for submissions.

10.15: Will paper (bubble) reports continue to be required for individual batch submissions or will alternate methods, such as EDI or spreadsheet be allowed for submissions?

A: No, as stated above in question 10.14, paper/bubble reports will no longer be required. Submissions will be in spreadsheet/flat file format.

10.16: Will EPA want electronic copies of quality control testing charts?

A: EPA will want the original records (including test records). The original charts can be provided electronically, or we will accept paper printouts of electronic copies (or other hard copies).

10.17: In the preamble, EPA states, "*the reporting forms can be standardized and the review process automated in such a fashion as to minimize the Agency resource requirements*" (page 39062). What is EPA's intent with regard to such standardization? According to EPA, commenters stated that the information needed for D & T is already kept as part of normal business practices (pages 39061, 39069). What is EPA's understanding of "normal business practices" and the way this information is already kept? How does this mesh with EPA's statement on standardization? What are EPA's expectations of D & T documentation? How will EPA address the different ways that facilities "already keep" the documentation? What is EPA's expectation with regard to PTDs? Is EPA looking to create or require a standard PTD? Will EPA request electronic copies of PTDs?

A: *We are proposing a flexible reporting system which historically has been well received by industry because of its ease of use. No special software will be required (reports should be submitted in spreadsheet or flat file format).*

10.18: What kind of local record keeping (at the terminal) will be required versus a centralized record keeping process?

  A: *Sections 80.592(e) and 80.602(e) require that the required records be made available "on request by EPA." The location where the records must be kept is not specified by the regulations. EPA interprets this requirement to require that records should normally be made available the same day they are requested. If they are stored off-site, EPA may request photocopies or facsimile copies be supplied pending receipt of original records, or EPA may visit the off-site facility to obtain the records. If voluminous records are requested by EPA, they must be provided within a reasonable time, which will generally be specified in the information request.*

10.19: Will the terminals or pipelines have to provide consignees to interconnecting pipelines any documentation, or can they forward documentation from the source, or is no documentation an option?

  A. *When a party transfers custody of designated fuels in the distribution system the transferor must provide a PTD to the new custodian. PTDs contain a variety of information, some of which will not be included in information generated by prior sources of the fuel. No documentation is not an option in cases where a PTD is required.*

10.20: Where (specific certified and express mailing addresses at EPA) should refiners and importers send the 2003, 2004 and 2005 pre-compliance information required in § 80.594?

  A: *Certified mailing address for pre-compliance reports is,*

  Attn: Diesel Sulfur (6406J)
  1200 Pennsylvania Avenue NW
  Washington, DC 20460

  Express (overnight/commercial carrier) mailing address for pre-compliance reports is,

  Attn: Diesel Sulfur Pre-compliance Report
  c/o: Chris McKenna (202-343-9037)
  1310 L St. NW
  Washington, DC 20005

Also, please send electronic copies of all documents, if possible, particularly the spreadsheet template that EPA developed for reporting estimates of 15 ppm production, 500 ppm production and credit generation and usage. Electronic copies can be submitted on a disk or e-mailed to mckenna.chris@epa.gov.

10.21: Not all on-road diesel fuel must be 15 ppm starting in 2006. The phase-in, small refiner, and geographic provisions of the final rule seem to assure that at least some 500 ppm diesel fuel will be in the market post-2006. But, for retailers and wholesale purchaser consumers, whether 500 ppm will be available in substantial quantities in their marketing areas is an important factor in deciding whether to invest in additional tankage in order to be able to offer customers both 15 ppm and 500 ppm diesel fuel. How can retailers and wholesale purchaser consumers gain information on the amount of 500 ppm that will be produced in their PADD post-2006?

A: *Refiners and importers are required to annually update EPA on their progress toward producing 15 ppm sulfur diesel fuel through the pre-compliance reports. We will produce a summary and analysis document of the pre-compliance reports submitted each year (the Summary and Analysis of both the 2003 and 2004 pre-compliance reports can be found at:* http://www.epa.gov/otaq/diesel.htm*). The annual summary and analysis reports will provide information, summarized and aggregated on a PADD basis, describing the volumes of 15 ppm and 500 ppm highway diesel planned to be produced, and estimates of the number of credits that refineries expect to generate or use. Further information on the business plans for individual parties in individual markets would need to be pursued directly by retailers and wholesale purchaser consumers.*

10.22: EPA released its status report on the on-road diesel sulfur rule earlier this year (2002). From a retailer's point of view, there wasn't much helpful information in this status report in terms of which refiners are going to make 500 ppm in 2006 and beyond and in what qualities. Will these reports by refiners to EPA provide estimates of 15 ppm diesel fuel production? Or of 500 ppm diesel fuel production?

A: *Refiners and importers are required to submit annual pre-compliance reports under the highway rule from 2003 through 2005, and under the nonroad rule as late as 2011. These reports must contain estimates of the volumes of 15 ppm sulfur fuel and 500 ppm sulfur fuel that will be produced at each refinery, and, for those refineries planning to participate in the trading program, a projection of how many credits will be generated or must be used by each refinery. These pre-compliance reports must also contain information outlining each refinery's timeline for compliance and provide information regarding engineering plans (e.g., design and construction), the status of obtaining any necessary permits, and capital commitments for making the necessary modifications to produce low sulfur highway diesel fuel. Much of this information will be claimed as confidential business information (CBI). As a result, we will be unable to release individual refiner plans unless they give permission to do so. However, we do report the information in an aggregated, non-confidential format. Refiner-specific information will have to be pursued through individual business relationships.*

10.23: What is the purpose of requiring record keeping and reporting for early compliance. If EPA is seeking to encourage early compliance, these requirements may be burdensome for certain marketers?

> A: *Record keeping and reporting for early compliance is required to ensure that 15 ppm credits are valid, as per § 80.531.*

10.24: Must terminal downstream hand-offs be "identical" to IRS reports, per § 80.601(d)(3)?

> A: *Previously, § 80.601(d)(3) required that the hand-off be identical, however this was changed in a recent technical amendment. The language has been revised to state that: "All reports shall be submitted on forms and following procedures specified by the Administrator, shall include a statement that volumes reported to the Administrator under this section are in substantial agreement to volumes reported to the Internal Revenue Service (and if these volumes are not in substantial agreement, an explanation must be included)..." The technical amendment was published in the Federal Register on November 22, 2005 and can be found at:*
> *http://a257.g.akamaitech.net/7/257/2422/01jan20051800/edocket.access.gpo.gov/2005/pdf/05-22807.pdf.*

11.   Testing

11.1:   Per § 80.604(d)(5), must aromatics be reported (and run) if compliance is based on cetane index?

A: *No. Under §§ 80.510 and 80.520, MVNRLM diesel fuel may comply with either the cetane index or the aromatics standard.*

11.2:   For testing purposes at a terminal level, must each tank that is filled be tested or is it sufficient to test just the incoming pipeline shipment?

A: *There are no testing requirements downstream of the refiner and importer to verify that the sulfur content fuels received by a distributor is consistent with that reported on the PTD. However, every party in the fuel distribution chain that has had custody of a fuel batch with a sulfur content found to be excess of the sulfur standard reported on the PTD could be held presumptively liable for the resulting violation if they lack sufficient affirmative defenses. The issue of whether testing the sulfur content of the incoming pipeline shipment rather than that of each storage tank into which the fuel is delivered to establish the sampling and testing element of a terminal owner's affirmative defenses will be evaluated by EPA on a case by case basis. In instances where the tanks being filled with 15 ppm diesel fuel have been dedicated to 15 ppm diesel use, it will be more likely that pipeline receipt testing would be sufficient and that a test on the tank after the introduction of the fuel would not be necessary. In cases where the tank is alternately used to store higher sulfur fuel as well as 15 ppm diesel fuel, a test on the tank after the introduction of the fuel may be needed.*

11.3:   Do terminals need to test every batch?

A: *Terminals are not required to test terminal tanks after every receipt of product. However, to establish a defense to presumptive liability to a violation, a terminal must establish each defense element. One such defense element is that the terminal had a quality assurance program, including a periodic sampling and testing program. Since terminals receive very large volumes of product in each receipt, an appropriate quality assurance program might require sampling and testing after each receipt of 15 ppm product. Note that terminals would have to test every batch in order to establish a defense if it is adding greater than 15 ppm sulfur content additives (or conduct the VAR approach under § 80.614, if the additive is a static dissipater additive).*

11.4:   Should testing at a terminal be done at the rack after injection of any additives?

A: *There are no sulfur testing requirements downstream of the refiner/importer. However, to meet their affirmative defense to presumptive liability, parties that blend >15 ppm sulfur additives into diesel fuel subject to the 15 ppm sulfur standard must conduct a periodic sampling and sulfur testing program after the additive is added to the fuel (see 40 CFR 80.613(d). Alternative affirmative defense requirements exist for blenders of >15 ppm anti-static additives that are based on a volume accounting reconciliation system (see 40 CFR 80.614).*

11.5: Will EPA be testing diesel for compliance at the retail/vehicle level?

A: *Yes, EPA will be testing diesel for compliance at the retail/vehicle level.*

11.6: Please elaborate on fuel tests at retail- what frequency would be recommended?

A: *Retailers are not required to perform testing in order to have a defense to presumptive liability. However, to establish a defense to a violation distributors must, among other things, conduct a periodic sampling and testing program. For a truck distributor, the best place to take samples may often be the tanks of retail outlets it delivers to. Retailers should probably sample and test after tank transition from a higher sulfur product to 15 ppm product, to confirm that the fuel in the tank is meeting the 15 ppm standard. Otherwise, if a violation is detected for some time subsequent to a tank transition, it may be difficult for the retailer to demonstrate that it did not cause the violation.*

11.7: When does the 2 ppm tolerance apply?

A: *The 2 ppm downstream sulfur test tolerance (§ 80.580(d)) applies at all locations downstream of the refiner/importer beginning on October 15, 2008. Prior to that date, a 3 ppm test tolerance adjustment applies at these locations.*

11.8: If XYZ Refining has a branded retailer who independently owns his/her XYZ Gas Station, does XYZ Refining have a duty to test ULSD at XYZ Gas Station?

A: *A branded refiner is not required to test gasoline at its branded retail outlets whether or not the retail businesses are owned by independent third persons. However, under § 80.612, branded refiners are liable for violations at branded retail outlets. Under § 80.613, for branded refiners to establish a defense to such violations, it must, among other things, show that it has a periodic sampling and testing program at the branded retail outlets.*

11.9: At what point in the distribution/receipt of ULSD will terminals need to sample/test for sulfur? If terminals do not test for sulfur content at the rack will they be able to assert an affirmative defense in the event that contaminated fuel is discovered at the retail level?

A: Under the diesel sulfur program, there is no one time or physical location that a distributor must utilize for sampling. The party must use its knowledge of the distribution system, and information regarding each receipt of product (e.g., whether the shipment may include fuel that is very close to an interface with higher sulfur product) to determine when and where to sample. However, as a general matter, we believe that a terminal should take samples during and/or after receipt of new product into the tank so that if there is a problem, it can be discovered and remedied before the product is distributed further downstream. If there is reason to believe that piping or a tank may contain high sulfur product then that may affect both sampling strategy and actions to prevent a potential violation. This is but one example of many possible situations. In regards to terminals not testing for sulfur content at the rack, if a terminal does not conduct a sampling and testing program, it will likely be difficult to be able to establish a defense to presumptive liability.

11.10: If tests are done upon receipt from the pipeline, how will PTDs reflect sulfur contamination added by the terminal through common piping, manifolds or at the rack?

A: These tests would not address contamination by the terminal. Also, PTDs do not need to specify the precise sulfur content of the fuel, only that the fuel is compliant with the applicable sulfur standard. If minor contamination occurs, but the fuel still meets the applicable standard, it is not considered a violation, nor does EPA require it to be reported on a PTD.

11.11: With regard to test methods, would EPA consider the following two approaches to be equally sound in terms of a defense to presumptive liability? (1) qualify the test method chosen for downstream oversight purposes per the precision and accuracy criteria in the rule; or, (2) correlate the test method chosen for downstream oversight purposes to a test method qualified per the performance based test method criteria? The methods appear to be equivalent and acceptable approaches.

A: No, for the testing of fuel for compliance with the 15 ppm sulfur standard, we do not consider the two approaches to be equivalent. The second approach listed is not an option for approval of a test method under the diesel fuel regulations, and therefore testing with methods not approved under § 80.581 and § 80.585 would not be considered an adequate defense to presumptive liability. Test results from test methods that have not been approved under the regulations might be considered as evidence of compliance or noncompliance with the standard, but use of the unapproved method would not be considered as meeting the periodic sampling and testing requirement for defense purposes; nor would use of an unapproved method be considered to have fulfilled the requirement for refiners and importers to test every batch. Testing for 500 ppm fuel has a different situation.

11.12: Is there a process, or a need, to certify a tank of 15 ppm sulfur content highway fuel that is not located in a refinery? Is it a single sample or Top-Middle-Bottom samples? Must it be by lab analysis?

> A: *Parties in the distribution chain downstream of the refiner or importer are not required to test motor vehicle diesel fuel. However, in order to establish a defense to any violation, downstream parties must conduct a quality assurance program, including sampling and testing. The fuel must meet the standard on a per-gallon basis. An appropriate field method may be used for downstream quality assurance testing. The regulations do not specify which testing method must be used for downstream quality assurance testing. The sampling methods prescribed by the rule in § 80.580 would be appropriate for purposes of quality assurance sampling under § 80.613(d)(1).*

11.13: Terminal operators believe it is appropriate for them, even when acting as a "downstream facility," to ensure that the diesel fuel leaving the terminal gate complies with the 15 ppm sulfur standard. Terminal operators recommend that EPA determine compliance when the product leaves the facility, not when the product is still in tank before distribution. However, if such testing and determination is to be conducted on product in storage, the EPA should use an "average" of the sulfur content of the tank to determine compliance.

> A: *Fuel represented to meet the 15 ppm standard that is stored, transported, dispensed, sold, offered for sale, supplied or offered for supply at a terminal must meet the 15 ppm sulfur standard on a per-gallon basis (not on an average basis). Terminals are not required to perform sampling and testing except as a defense to liability, but we strongly concur that responsible terminal operators will conduct regular sampling and testing – and that the appropriate frequency will probably be after each receipt of product.*

11.14: Will EPA have an oversight testing program for 15 ppm motor vehicle diesel fuel similar to the RFG gasoline survey program?

> A: *The RFG survey program is a statistically valid sampling and testing program that is performed by an independent contractor who is paid for by participating refiners. Its purpose is to show whether the RFG in each control area meets standards on average. No such program exists under the highway or nonroad diesel regulations. However, as with other fuels programs, EPA plans to conduct its own sampling and testing at all levels of the distribution system to monitor compliance with the diesel fuel standards.*

11.15: As EPA is aware, terminals often serve in two different capacities – (1) as an import facility; and (2) a downstream facility when product is received from a U.S. refiner or importer. As such, terminal operators believe that the 2 ppm testing tolerance would only apply to those operations of the terminal when it is acting as a "downstream facility." Would EPA please confirm this interpretation.

> *A:*   *The 2 ppm test result adjustment of § 80.580 (d) does not apply for fuel at a facility that is acting in the capacity of an import facility. However, fuel that is in a terminal tank downstream from the refinery or import facility is eligible for the 2 ppm test result adjustment. (Please note that, per our May 2006 technical amendments to the regulations, prior to October 15, 2008, this downstream test tolerance adjustment factor is 3 ppm.)*

11.16: Does other testing need to be done to ship No. 1 diesel and redesignate it as jet to get ULSD credit?

> *A:*   *No other testing would need to be done. However, the anti-downgrading provisions would apply if a downgrade (to 500 ppm highway diesel) were to occur. The situation that the question is asking about would not affect the person's volume balance; if the reverse were to occur- jet fuel was redesignated as No. 1 diesel fuel- the volume balance would be affected.*

11.17: Are there regulations that <u>require</u> sampling and testing by <u>each</u> facility, or do you just need a sampling and testing program to establish a defense?

> *A:*   *In general, the only parties that must sample and test are refiners and importers; under § 80.581, they must sample and test each batch of motor vehicle or NRLM fuel subject to the 15 ppm sulfur standard that is produced or imported. However, a downstream party redesignating kerosene fuel for use as 15 ppm motor vehicle or NRLM fuel when such fuel is not accompanied by a PTD stating that it meets the 15 ppm standard, must have test results (see § 80.525). Likewise, any party remedying a contamination of fuel originally designated as 15 ppm fuel, or upgrading fuel not designated as 15 ppm fuel to fuel that meets the 15 ppm standard must sample and test before distributing the fuel as fuel that meets the 15 ppm standard.*

12.   Enforcement

12.1:   What will happen if you refuel a 2007 truck with contaminated fuel? Are there any engine or aftertreatment issues? If a 2007 truck is misfueled, is there any legal liability for the end-user?

A: *If a MY 2007 or later truck is misfueled <u>once</u>, it will have significantly higher PM emissions during operation on that fuel, but there should not be any significant long-term emissions or engine durability concerns as long as the truck is then fueled with the proper fuel. A wholesale purchaser-consumer or end user faces presumptive liability if it misfuels vehicles or equipment with fuel not meeting applicable standards. However, a vehicle operator fueling at a retail outlet from an appropriately marked pump is not liable for the misfueling caused by the retailer or other party. A wholesale purchaser-consumer (fleet operator with its own fueling facility– see definition at 40 C.F.R. § 80.2) may be able to establish a defense if the fuel was contaminated by an upstream party and the wholesale purchaser-consumer can otherwise establish in its defense elements that it did not cause the violation and that product transfer documents account for the product and demonstrate the product was compliant.*

12.2:   EPA appears to focus enforcement on the custodian of the fuel, even though a compliance violation may well have occurred further upstream. Will EPA automatically require custodians to raise a defense and overcome a presumption of liability for simply possessing off-spec fuel?

A: *The facility where the violation is found and all parties in the distribution system upstream of that facility will be presumed liable. However, each party has the opportunity to establish a defense to the presumptive liability. For example, an upstream distributor who takes custody of a fuel would be presumed liable, and would need to demonstrate that it did not cause the violation- that product transfer documents account for the fuel and show that the fuel was apparently in compliance when it was in its custody and that it has an adequate periodic sampling and testing program (regardless of whether or not it tested that specific batch of fuel). However, if the fuel custodian is a retailer or wholesale purchaser-consumer, they would not be required to have a periodic sampling and testing program in order to establish a defense.*

12.3:   The regulation requires an entity to conduct an investigation into the cause of a quality control testing violation. What evidence will EPA require for purposes of establishing that an adequate investigation was conducted?

A: *EPA is not in a position to discuss what may constitute an adequate investigation in every conceivable scenario. An adequate investigation is whatever is required to correct the cause of the deviation and to assure that the method performance has been restored to statistical control.*

12.4: The rule states that a party that is initially deemed liable for a violation of the rule will not be deemed liable if the party demonstrates: (1) the violation was not caused by the party; (2) product transfer documents show that the violating product was in compliance when it was under the party's control; (3) the party conducted a quality assurance sampling and testing program. The regulations, at § 80.613, do not specifically require a party to prove all three factors (there is no "and"). However, the preamble (page 39104) says that a party must show 1, 2 "AND" 3. It should be enough to show that the party did not cause the violation (first prong). How will EPA apply this test?

   A: *The preamble is correct- a party that is presumed liable must establish all three elements (except for retailers and wholesale purchaser-consumers, which are not required to have a periodic sampling and testing program). This is consistent with all of the fuels rules in Part 80, and was corrected in a recent technical amendment (July 7, 2005). For example, where a truck distributor drops 500 ppm sulfur content fuel into a retailer storage tank that supplies a 15 ppm pump stand, both parties will be presumed liable. However, the retailer may establish a defense- even if it subsequently sold the fuel as 15 ppm fuel- if it can demonstrate that it did not cause the violation and that the PTDs account for the fuel as being 15 ppm fuel, and thus show the fuel to be in violation. (In this example, if the party in question was not a retailer, it would also need to demonstrate- regardless of whether or not it tested that specific batch of fuel- that it has an adequate sampling and testing program in place). If the product transfer document stated that the fuel was 500 ppm fuel, then the retailer cannot establish a defense unless it immediately stopped sale and took actions (probably in association with the distributor) to remedy the violation. If the retailer sells the product as 15 ppm fuel even though the product transfer document shows 500 ppm fuel was dropped, then it cannot meet the second element of its defense.*

12.5: Under the rule, sampling and testing is voluntary for downstream parties, however, to establish a defense to presumptive liability, they will engage in periodic testing. If a downstream party is not in the midst of periodic testing and is found in possession of an off-spec batch, then the party will not be able to meet all of the three elements required for establishing a defense (see question 12.4, above). It would appear that it might be impossible to rebut the presumption if a party is not in the midst of scheduled QC sampling and testing. Is this correct?

   A: *If a party conducts no quality assurance/sampling and testing, it cannot establish a defense. However, a party can satisfy the periodic sampling and testing element of its defense by demonstrating that its periodic sampling is done at a frequency that is appropriate given all of the circumstances. It is not necessarily required that the product EPA finds to be in violation was actually sampled and tested. However, for parties such as terminals, an adequate sampling frequency may require sampling after each receipt of product. EPA will investigate whether other parties may be liable, but each party must be able to establish a defense.*

12.6: The end user has no way of knowing whether they are refueling with diesel in excess of the 15 ppm sulfur regulatory limit. Is there a way to introduce some flexibility in the enforcement of the requirement that no fuel in excess of 15 ppm sulfur be introduced into a 2007 (or later model year) truck?

A: *Flexibility inherently exists in the fact that everyone involved in the production and distribution chain of each batch of fuel is presumptively liable, which will help to limit the occurrence of fuel that does not meet the 15 ppm standard (or other applicable designation).*

A person fueling a vehicle at a retail outlet self-serve pump will not be liable for misfueling if the person does not cause the violation. For example, if the misfueling occurs because the retail outlet has mislabeled the pump stand or because the retailer or its distributor have contaminated the fuel with high sulfur product, the retailer and parties upstream may be liable, but the self-serve customer will not be liable unless it played some part in causing the violation. However, if, for example, a person fuels a 2007 MY truck at a self serve pump labeled as dispensing 500 ppm fuel, the end user is liable.

12.7: Will testing documentation from a receiving pipeline be sufficient evidence of compliance with the sulfur standard to continue delivery? Will this document track through to a truck load without additional testing and be sufficient to establish a defense to liability for a violation of the sulfur standard?

A: *EPA will sample and test diesel fuel at all levels of the distribution system. If, for example, the sulfur test result for a particular sample downstream of the refinery exceeds the 15 ppm sulfur standard (after the adjustment under 40 C.F.R. § 80.580(d)), that fuel will be in violation regardless of any prior test result demonstrating compliance. Note that fuel from each refiner and each pipeline is frequently commingled in the fungible distribution system. Fuel may be contaminated in the pipeline or in storage tanks during shipment.*

Where a violation is found, the party having custody of the fuel that is found in violation, and any party upstream of the facility where the violation is found is presumptively liable. However, each party may establish a defense by meeting each of its defense elements. For example, a distributor may establish a defense to a violation found at a retail outlet if it can show that: it did not cause the violation; product transfer documents account for the fuel and indicate that the violating product was in compliance with applicable requirements when it was under the party's control; and the party conducted a periodic sampling and testing program and other quality assurance efforts. Under § 80.613(d)(4), truck carriers can rely on sampling and testing conducted by another party, but must make other appropriate quality assurance efforts to meet this defense element. Parties other than tank truck carriers will need to conduct their own periodic sampling and testing, and will not be able to rely in whole on testing by parties upstream in the distribution system.

12.8: Will appropriate processes and QA/QC oversight be acceptable defenses against enforcement for off-spec product? Testing shows that the pipeline does not add sulfur from the steel. If base testing is done to identify protection volumes, and operational records support no accidental contamination, will this be sufficient? Could this preclude many proprietary and batched pipelines from needing on-line analyzers and extensive sampling process?

A: *Parties are presumed liable for a violation if a noncompliant fuel is found to be in their control, or if they are in the distribution system upstream of the facility where the violation is found. Downstream parties have a defense to presumptive liability if: (1) the party did not cause the violation; (2) PTDs establish that the fuel was in compliance while the fuel was under the party's control; and (3) the party conducted an adequate QA program, including sampling and testing.*

While refiners, importers and transmix processors are required to test each batch of product represented to meet the 15 ppm sulfur content standard, most downstream parties are not required to test every batch. Testing for downstream parties is voluntary. Downstream parties may conduct periodic sampling and testing for quality assurance purposes in order to establish a defense to liability for alleged violations.

Each company should consider its distribution system and its own operations in determining the appropriate sampling and testing frequency. Factors that EPA believes are relevant include: the results of previous sampling; the volume of fuel in a particular shipment (the larger the volume, the greater the justification for testing); the degree of confidence in the quality of the product when it was received; and the opportunity for violations while the fuel was in possession of the party (e.g., the opportunity for commingling with higher sulfur product). For example, the volume of shipments into terminals are normally relatively large. A terminal that samples and tests after each receipt of 15 ppm sulfur content product will likely be able to meet its periodic sampling and testing defense element, and will have significant evidence regarding the "didn't cause" defense element as well. However, it should be noted that a terminal could demonstrate that all product received meets the applicable specifications, and still cause a violation to occur subsequent to receipt. Also note that there are variations to the defense elements for certain parties, such as parties who blend additives having a sulfur content exceeding 15 ppm to fuel subject to the 15 ppm standard.

Note that the rule requires QC for sulfur measurement instrumentation. If instruments are out of control requirements, then all samples taken since the instrument was last in control must be retested. QC records must be retained for 5 years.

12.9: Will pipelines have an exempt refiner defense if higher than standard sulfur is located in their systems?

A: *The fuel sulfur level of a batch of fuel must conform to the way it is designated. If a pipeline has fuel that is designated as 500 ppm or 15 ppm sulfur, and it does not meet the designated level, then it is noncomplying. It is not a defense to claim the fuel should have been designated otherwise by an upstream party.*

12.10: What steps must a retailer/distributor take to assure that it is not liable for the sale of off-spec motor vehicle diesel fuel subject to the 15 ppm standard?

A: *As discussed in questions 4.35 and 12.18, retailers have certain affirmative requirements, including properly labeling pump stands, maintaining product transfer documents and records regarding downgrading of 15 ppm diesel fuel. Retailers must not cause the 15 ppm fuel to be contaminated by misdelivery of high sulfur product into a storage tank for 15 ppm sulfur content product. Retailers should carefully inspect product transfer documents at the time of delivery to assure that the proper product is being delivered into each storage tank.*

We believe that retailers and distributors should work closely together to assure that misdeliveries do not occur.

The liability and defense provisions of the 15 ppm motor vehicle diesel rule are similar to those of previous fuels rules. Under § 80.613(a), a retailer and any distributor or carrier in the distribution chain are deemed liable for a violation found at the retail outlet (see § 80.612). A retailer will not be deemed in violation if it can show that it did not cause the violation, and product transfer documents account for the fuel found to be in violation and indicate the violating product was in compliance with the 15 ppm sulfur standard when it was under the retailer's control.

A distributor must also show that it did not cause the violation, that the product transfer documents account for the product, and show that the product was in compliance when it was under the distributor's control. A distributor could cause a violation by various actions, including by misdelivery of high sulfur product or commingling high sulfur product with 15 ppm product in a storage tank or transport truck. To establish a defense, a distributor must also demonstrate that it has conducted a quality assurance program, including periodic sampling and testing of the fuel it transports and delivers (see § 80.613(d)).

12.11: Many retailer/distributors use common carriers, rather than their own trucks, to transport and deliver diesel. Under the diesel rule, what potential liability and defenses to liability are in place for common carriers? If a common carrier does not follow the supplier's or the retailer's location and/or delivery instructions, is the common carrier liable for the misdelivery or the retailer?

    A: *As with other EPA fuels programs, the highway and nonroad diesel sulfur regulations utilize a presumptive liability structure. When a violation is found, the party who owns, leases, operates, supervises or controls the facility where the violation is found, and every party upstream of that facility, who supplied the fuel to the facility where the violation is found, is presumed liable, absent a complete defense to the violation.*

        Under the hypothetical, initially both the carrier and the retailer would be deemed in violation. The carrier who delivers 500 ppm product to a retail outlet tank when 15 ppm product has been ordered would not be able to establish it did not cause the violation, and thus would remain liable.

        The distributor who hired the common carrier would be liable if the product transfer documents and other paperwork demonstrated that 15 ppm product was ordered and 500 ppm product was delivered, and the distributor failed to take reasonable steps to address the situation.

        A retailer may establish its defense by 1) demonstrating it did not cause the violation; and 2) and demonstrating that product transfer documents account for the fuel found to be in violation and indicate that the violating product was in compliance with the standards. A retailer does not have to conduct a quality assurance sampling and testing program to meet its defense requirements. However, a retailer must check the product transfer documents and properly direct deliveries to the proper storage tank. In this hypothetical, the retailer may have been able to detect the misdelivery at the time of delivery if it had checked the PTD provided by the carrier. If the retailer does not check the PTD at the time the misdelivery occurs (and if doing so would have shown that a misdelivery had occurred), and sells the product as 15 ppm fuel, the retailer would not be able to establish a defense.

12.12: How about after-hours deliveries? Many retailers also are not open 24 hours a day and received diesel deliveries during the night while its stores are closed. If a misdelivery is made, what are the retailer's defenses?

A: *As discussed above, regardless of when the delivery occurs, the retailer must properly direct deliveries, and it must check the product transfer documents. Where the retailer's employees will not be present during deliveries, both the retailer and the trucker should take special care to prevent misdeliveries. A product transfer document must be provided to the retailer with every transfer of custody or title of the diesel fuel. The retailer has the responsibility to check the transfer document to determine that fuel meeting the appropriate standards was delivered to the appropriate tank. Therefore, if a misdelivery is made after-hours, the retailer's employee should discover this misdelivery upon opening the outlet the following day through an inspection of the product transfer document left by the carrier (or possibly earlier if a transfer documents were received earlier). If the retailer did not cause the misdelivery, and it discovers the misdelivery through inspection of product transfer documents, and locks the pumps before any diesel fuel is sold, then it will not be held liable for the violation.*

Assuming the retailer properly instructed the distributor or carrier to deliver 15 ppm fuel to the appropriate storage tank, the carrier would not be able to establish a defense. The distributor would also be liable for a violation if the paperwork shows that a misdelivery was taking place (e.g., that 15 ppm fuel was ordered but 500 ppm fuel was being delivered).

12.13: Refiners will produce diesel fuel that meets the 15 ppm standard at the refinery gate. However, the rule mandates that the sulfur content remain at 15 ppm or below throughout the distribution system until it is dispensed into a vehicle's fuel tank. Assume the PTDs of the product are in order and properly identify the diesel as 15 ppm fuel, what potential liability could a retailer have if the batch has been contaminated -- in the pipeline, at the terminal, or in the transport truck -- so that the diesel received by the retailer exceeds the 15 ppm spec?

A: *In this case, a retailer would initially be presumed liable for the violation, but could establish a complete defense to liability if the PTDs account for the fuel in violation and show that the product was in compliance with the 15 ppm sulfur standard; and the retailer did not cause the contamination. There may be cases where an unscrupulous retailer purposely receives violating product delivered by "midnight drops." In some cases no product transfer documents cover the product. The retailer in such cases would not be able to establish a defense, since it would have caused the violations, and since not all product would be covered by its product transfer documents.*

Once the retailer becomes aware that the sulfur content of diesel product subject to the 15 ppm standard exceeds the standard, it has an obligation to shut down its retail pumps affected by the contaminated product until the violation can be corrected. If the retailer did not shut it off, it would be liable for selling or dispensing noncomplying diesel fuel.

12.14: There is no field testing equipment commercially available to measure accurately sulfur at these low levels and almost no retailers or wholesale purchaser consumers have the type of on-site laboratories to conduct the testing. Thus, a distributor conducting a quality assurance plan (QAP) would be forced to send a sample to an outside laboratory for analysis and may not receive the test results for 24 to 48 hours. By this point, most if not all of the off-spec fuel would have been sold by the retail outlets, making it impossible to correct the non-compliance. Shouldn't a retailers or wholesale purchaser consumers QAP under the 15 ppm motor vehicle diesel program be limited to examining PTDs? If not, how does EPA expect distributors to comply with this defense requirement?

> A: *Retailers are not required to show an appropriate sampling and testing program to establish a defense to liability. (See § 80.613(a)(1)(iii).) However, distributors, including truck distributors, who take title to the fuel and deliver product to their retail outlets, must conduct a quality assurance sampling and testing program. These requirements are the same as in other fuels rules. A truck distributor could rely, in part, on a proper quality assurance sampling and testing program conducted by the terminal. But since there may be opportunities for the trucker to obtain improper product (e.g., 500 ppm fuel instead of the 15 ppm fuel that was ordered), or commingle high sulfur products with 15 ppm product, or otherwise cause violations, sampling at locations downstream from the truck loading terminal would be appropriate.*
>
> *If a distributor must utilize an outside laboratory to obtain test results, and if it is impossible for the distributor to obtain results quickly, the periodic sampling and testing quality assurance program would still be useful in as much as it would show if any violations are occurring and based on those findings the distributor and retailer should act immediately to prevent future violations, and to remedy the present violation if it still possible to do so by shutting down the affected pump stands until the product meets standards. In this situation, if a violation occurred and is discovered through the quality assurance testing, the distributor and retailer may not be able to establish a defense to the violation based on the delayed test results. However, EPA may take any such unavoidable delay into account in its evaluation of the case. Moreover, the distributor could, as stated above, limit its exposure, by obtaining as many terminal test results as possible. Whether the distributor and retailer could meet the causation element and PTD element would depend on the specific facts.*

12.15: What responsibility, if any, does a retailer have to "police" its customers to assure that they do not misfuel their vehicles -- e.g., dispense 500 ppm diesel, if offered at the retail outlet, into a 2007 or newer heavy duty truck? In many cases, diesel retail outlets are large with multiple fueling locations and are staffed by entry level employees earning just over the minimum wage. Does EPA expect these employees to "police" customers at the retail outlet to assure that these customers are fueling properly? Also, please provide further clarification of the term "permit." at § 80.610(d)(1). For example, if a marketer tells a self service customer that he can't fuel with 500 ppm but the customer continues, is the marketer "permitting" the introduction of 500 ppm into a 2007 or newer model year vehicle requiring 15 ppm? When exactly is a marketer "permitting" misfueling?

A: *Under the regulations, no person shall "[i]ntroduce, or permit the introduction of, diesel fuel into model year 2007 or later diesel motor vehicles, and beginning December 1, 2010 into any diesel motor vehicle, which does not comply with the standards and dye requirements of § 80.520(a)and (b)." Nor shall any person "introduce or permit the introduction into model year 2007 vehicles, motor vehicle diesel fuel that is identified as other than diesel fuel complying with the 15 ppm sulfur standard...." Nor shall any person cause another person to commit an act in violation of these prohibitions. 40 CFR § 80.610(d) and (e).*

Therefore, a retailer will be liable if it, or its employees, introduce noncomplying fuel into vehicles, or permit such introduction of noncomplying fuel into vehicles or cause another person to violate the stated prohibitions. A retailer must properly label pump stands and assure that product represented by product transfer documents as meeting the 15 ppm standard is delivered to the appropriate storage tanks, and that product subject to the 500 ppm sulfur standard is not delivered to such storage tanks. Moreover, a retailer should train its employees that may fuel diesel vehicles regarding these prohibitions and the pump stand labels and the corresponding labeling of filler inlets and dashboards on model year 2007 and later diesel vehicles. Because there may be many scenarios regarding how misfueling violations occur at retail facilities, a retailer's liability for misfueling that occurs at its facilities will be evaluated on a case-by-case basis.

Where a retailer has made reasonable efforts to prevent self-service customers from misfueling (e.g., by cutting power to the pump as soon as the misfueling is perceived) and has not in any way caused the misfueling (e.g., through mislabeling pumps) the retailer would not be liable for "permitting" misfueling by a self-service customer.

12.16: What about intentional misfueling by a customer? In the past, as new fuels have been introduced in different parts of the country (CARB II Diesel in California; RFG in Milwaukee), negative reactions by customers have been significant. Given the fact that truckers may believe that use of 15 ppm diesel fuel will result in lower fuel economy, less power, or mechanical problems with engine seals, some portion of customers might refuse to purchase 15 ppm fuel for their new trucks. How can EPA expect a retailer to stop a customer from intentionally misfueling his or her 2007 truck?

A: *A retailer is not responsible for misfueling by the customer unless the retailer or its employees have introduced the fuel, or permitted or caused the violation.*

While intentional misfueling by a customer normally does not meet this criteria, there may be scenarios where retail outlet employees' actions or failure to act may permit or cause violations, even though the vehicle operator may actually dispense the fuel into the vehicle.

We believe that responsible retailers will take reasonable actions to make misfueling less likely to occur at their retail outlets. For example, a retail employee could remotely shut off the fuel supply to a customer that it does see misfueling, and explain to the customer that the retailer cannot permit misfueling of vehicles at its stations since the retailer is potentially liable for such violations.

12.17: What about intentional misfueling by a fleet customer? For example, if a distributor's fleet customer orders a load of 500 ppm on-road diesel fuel, which the distributor delivers, and then the fleet customer intentionally fuels MY 2007 vehicles with the 500 ppm diesel, what is the distributor's potential liability? Does the distributor have the responsibility to "police" its fleet customers to assure there is no intentional misfueling?

A: *A distributor is not responsible for misfueling by fleet customers unless the retailer or its employees have introduced the fuel, or permitted or caused the violation. Intentional misfueling by a fleet customer normally does not meet this criteria. However, there may be scenarios where a distributor could be liable for permitting or causing misfueling violations by a fleet operator. For example a distributor who knows a fleet customer only has one diesel fuel storage tank, and knows that the fleet operator has model year 2007 or later vehicles, may be liable for permitting or causing misfueling if it delivers 500 ppm product to that customer's storage tank, even if the customer ordered the 500 ppm product. We believe distributors and carriers should work with retail and fleet customers to ensure that fuel is delivered to the appropriate storage tank.*

By "permitting" we mean that a retail station does not take reasonable actions to stop the intentional misfueling that it witnesses a customer perform.

12.18: With the 15 ppm maximum sulfur requirement (with the 2 ppm downstream testing adjustment under § 80.580(d)) applying at all levels of the distribution system, from refinery to retail outlets, how much service station, delivery truck, terminal testing and other quality assurance testing does EPA view as necessary to constitute a presumptive liability defense?

A: *A quality assurance program, including a sampling and testing program, is one of the defense elements under the presumptive liability scheme (see § 80.613(a)(1) and (d)). The liability and defense provisions under this rule are like those of other fuels rules. What constitutes a valid sampling and testing program varies depending on the type of party, as discussed below, and also varies depending on the particular circumstances involved. Factors that EPA believes are relevant in designing any sampling and testing program include: the results of previous sampling and testing; the volume of fuel being handled by a facility and the volume of individual batches being processed by a facility (the greater the volume of a batch the greater the justification for sampling and testing that batch); the degree of confidence in the quality of the fuel received based on all circumstances; the opportunity for violations to occur while the fuel is in possession of the party (e.g., the opportunity for commingling of 15 ppm fuel with 500 ppm or high sulfur distillates); and the opportunity for misdeliveries of product.*

Retailers and wholesale purchaser-consumers-
Retailers and wholesale purchaser-consumers are not required to conduct sampling and testing programs. See § 80.613(a)(1)(iii). However, during the period when storage tanks are being turned over from 500 ppm fuel to 15 ppm fuel, retailers should take steps to assure the product in the tanks meets the 15 ppm standard before making the fuel available for sale as 15 ppm fuel.

Terminals and Pipelines-
An acceptable periodic sampling and testing program for terminals will probably require sampling and testing after every receipt of product into a terminal storage tank, given the stringent sulfur standard, the potential substantial emissions consequences of violations, and the potential for contamination of product with high sulfur distillates (and especially in the early years of the program, high sulfur gasoline). Appropriate pipeline testing will depend on many factors, including the length of the pipeline, the nature of the product in the pipeline cycle on either side of the 15 ppm motor vehicle diesel fuel, the opportunities for contamination as product enters the pipeline or offloaded to terminals, etc. Carriers such as pipelines may, in addition to appropriate testing of their own, rely on testing by other parties, such as the refiners who deliver product to the pipeline.

Distributors-
Distributors of motor vehicle diesel fuel who take title to the fuel must conduct a periodic sampling and testing program as part of the quality assurance element of their defense to presumptive liability. The appropriate frequency of sampling would depend on the circumstances, as noted above. A truck distributor who delivers fuel to retail outlets may be able to rely on tests conducted by the terminal it loaded product from, if the sampling and testing program is properly performed, in order to show fuel picked up from the truck loading terminal met standards. However, if for example, the distributor delivers both 500 ppm product and 15 ppm product, periodic sampling at the distributor's facility (especially if the distributor has storage tanks) and at retail outlets it delivers to would be important. In addition, a truck distributor should employee procedures

106

to prevent misdelivery of 500 ppm product into 15 ppm product storage tanks, and to prevent contamination in delivery tanks.

Truck Carriers-
Under § 80.613(d)(4), a tank truck carrier may conduct certain other oversight activities in lieu of sampling and testing. Truckers should develop procedures to assure that 15 ppm sulfur content product they handle is not contaminated by higher sulfur products and to assure that 500 ppm fuel, or other high sulfur product, is not misdelivered into a storage tank for 15 ppm fuel. Truckers should have procedures to assure that high sulfur product is completely drained from a truck compartment and hoses before that truck compartment is used to carry 15 ppm motor vehicle fuel. Alternative oversight activities in lieu of testing include driver training; periodic review of records; providing drivers with specific information regarding which customers, and which storage tanks at customer's facilities, require fuel meeting the 15 ppm standard and which have 500 ppm fuel storage tanks (or nonroad storage tanks), etc. Truck carriers may also rely on the sampling and testing programs of other parties as part of their quality assurance program (e.g., sampling and testing conducted by their immediate supplier).

Branded Refiners-
A branded refiner must conduct a quality assurance sampling and testing program at branded retail outlets as one aspect of the quality assurance program defense element to branded refiner liability under § 80.612(a)(3). The branded refiner must assess the specific factors that apply to its branded retail outlets and the distribution systems that supply them in determining the appropriate sampling and testing rate.

In addition to the factors listed above, branded refiners should also consider the historical compliance of branded facilities in a marketing area, and whether there is reason to believe that particular downstream facilities do not comply with contractually imposed requirements designed to prevent violations. Moreover, in the first year the 15 ppm standard is in effect, a relatively high sampling rate would probably be necessary to assure that retail tanks are successfully transitioned to 15 ppm fuel by October 15, 2006. In subsequent years, it may be appropriate to raise or lower this sampling rate, depending on the extent of compliance demonstrated by the program. In areas where problems are found, the sampling rate should be increased, and efforts should be made to discover the causes of the problems and correct those problems. The specific facilities where violations were found should be reinspected frequently.

Refiners and Importers-

107

Under § 80.613(a)(1)(iv), a refiner or importer must conduct testing of every batch to meet the quality assurance periodic sampling and testing element of its defense to presumptive liability.

See § 80.613(a)(1)(iv) and (v), and § 80.613(d)(2) for special situations.

12.19: Marketers frequently have disagreed with EPA actions regarding waivers or enforcement discretion with respect to the RFG program. Some have predicted on-road diesel supply shortages when this rule takes effect in 2006. If such supply disruptions -- whether caused by underproduction; refinery, terminal, or pipeline problems; or, weather -- do occur, what will EPA's reaction be?

> A: *It is difficult to discuss possible enforcement discretion issues that may occur. However, we note that, especially after 2010, when all highway diesel fuel must meet the 15 ppm sulfur standard, we do not foresee any practical way to distribute motor vehicle diesel fuel having a sulfur content greater than 15 ppm into the motor vehicle diesel fuel market.*
>
> *Of course, we realize that temporary supply disruptions may occur due to an Act of God, or other unforeseeable event. The regulation, and the Agency's inherent discretion, provide mechanisms for EPA to consider whether a supply disruption is one for which relief is appropriate. However, we believe it would be very difficult now to address a variety of crises that may or may not occur in the future.*

12.20: There will likely be significant adjustments for the diesel fuel distribution system to make to assure that 15 ppm highway diesel is available on-spec for truckers and other customers. The rule mandates that this fuel be distributed in mid-2006 despite the fact that the demand for this fuel will be sparse prior to 2007. Would EPA consider providing a period of enforcement discretion after mid-2006 for companies in the distribution chain if good faith efforts to meet the 15 ppm standard are made?

> A: *The regulations created a 3-step implementation process that is expected to give all levels of the distribution system, including the retail outlet and wholesale purchaser-consumer level, enough time to bring storage tanks into compliance in time to meet the regulatory deadlines. Refiners and importers must be in compliance with the 15 ppm standard by June 1, 2006. All facilities downstream of the refiner or importer, except retail and wholesale purchaser-consumer facilities, must be in compliance by September 1, 2006, if they choose to market 15 ppm fuel. Similarly, retailers and wholesale purchaser-consumers must be in compliance by October 15, 2006, if they choose to market 15 ppm fuel. These entities downstream of the refinery have built-in flexibility in that they can continue to market 500 ppm fuel if they so choose until 2010.*

12.21: With the recent buying and selling of refining and marketing assets, branded retail outlets may no longer be supplied by refineries with the same name. E.g., Refiner A may now have retail outlets with Refiner B's brand and Refiner A may not supply the outlets. Given this situation, EPA's branded retail defense requirements no longer seem appropriate.

> A: *Branded refiner/importer liability under the regulations is not based on whether the branded refiner actually supplies the fuel to the retail outlet. Under § 80.612, any refiner or importer whose corporate, trade or brand name, or whose marketing subsidiary's corporate, trade or brand name appears at a facility where a violation occurs, is deemed liable for the violation and must establish its defense to branded refiner liability to avoid liability.*

12.22: In the fuel marketing industry "branded" means an exclusive supply contract between the brand and the retail outlet. In this sense some stations are branded for gasoline but not for diesel even though the "brand sign" flies above the property. Does branded refinery liability apply to diesel fuel sold at such a station?

> A: *Under § 80.612, if the brand name "appeared at a facility" where a violation occurred, the branded refiner would be liable under the branded refiner liability provisions. Where gasoline and diesel fuel are sold at a retail outlet on the same premises, and the facility is branded as defined by the regulation, the gasoline and diesel operations would generally be considered branded since they would usually be part of the same facility.*

12.23: Where a branded jobber owns & supplies its branded retail outlet, who is potentially liable for violations found at the retail outlet and what quality assurance sampling and testing is the branded refiner and jobber/retailer responsible for?

> A: *The branded refiner would be liable for any violation that occurs at the branded facility. To establish a branded refiner defense, the branded refiner must, among other things, establish that it had an appropriate quality assurance sampling and testing program. See questions 12.18 and 4.35 for liability and defense provisions for retailers and distributors.*

12.24: For an owner of a retail outlet or a fleet fueling facility (a wholesale purchaser-consumer facility), if the product transfer documents account for all product in the storage tanks and show that the product is in compliance with the 15 ppm standard, am I still at risk for penalties?

> A: *Where PTDs account for all product and show the product is in compliance with the diesel standards, and where the pumps are properly labeled, and the retailer or wholesale purchaser-consumer has not caused the violation, (e.g., has not misdirected noncomplying product into the 15 ppm diesel fuel storage tank or*

otherwise caused the contamination or misdelivery to occur) it will generally be able to establish its defense to liability. However, if for example, a retailer or wholesale purchaser-consumer were to purchase fuel that it has reason to believe does not comply with the 15 ppm sulfur standard, even though the PTDs account for all product and state that the product does meet the standard, the retailer would be liable since it would have caused the violation.

12.25: During the lead phasedown in gasoline, EPA mandated that larger nozzles be installed on gasoline dispensers with leaded fuel and that restrictor plates be installed on the fill pipes of vehicles' fuel tanks, preventing the fueling of cars designed for unleaded gasoline with unleaded gasoline. Is EPA considering a similar alteration in nozzle sizes to prevent accidental or intentional misfueling under the 15 ppm diesel fuel program?

A: *We did not finalize any provisions beyond fuel pump labeling requirements. We recognized that some potential for misfueling would still exist and continued to discuss options in meetings and workshops with industry to explore simple, cost-effective approaches that could further minimize misfueling potential. Through these discussions, however, no consensus was developed such that the magnitude of the potential misfueling problem would justify the cost and burden associated with additional regulatory controls.*

12.26: For presumptive liability it has been indicated that EPA will look along the chain of hand-offs; however, in the case of misfueling, will pipeline PTDs be required to establish liability, or for upstream parties to establish a defense?

A: *Where a retailer's or wholesale purchaser-consumer's fuel pump is properly labeled as dispensing 500 ppm product, but the fuel is dispensed into a vehicle requiring 15 ppm fuel, EPA would not take enforcement action against any person upstream of the retail outlet or wholesale purchaser-consumer unless such party caused the violation. Therefore, EPA could establish liability for the appropriate parties without inspecting pipeline PTDs.*

If the misfueling was caused because a fuel pump labeled as dispensing 15 ppm fuel actually dispensed fuel with a sulfur content exceeding that standard, then presumptive liability would apply to the retailer or wholesale purchaser-consumer and to all upstream parties (and to the branded refiner, if applicable). For each of these parties, one of its defense elements is that it demonstrate that it has PTDs that account for the product and show it was in compliance when in the possession of that party.

12.27: Assume a terminal tank contains diesel fuel meeting the 15 ppm sulfur standard and that fuel is contaminated by a delivery of off-spec diesel fuel represented to meet the 15 ppm standard such that the terminal's test results after receipt show that the sulfur content of diesel fuel in the terminal tank is now greater than 17 ppm. What are the consequences and what options are available to the terminal?

A: *Finding that fuel represented to meet the 15 ppm standard exceeds 17 ppm at a downstream location such as a terminal has consequences regarding compliance with both the standards and with the "designate and track" provisions.*

Liability and Remedial Action
Under the presumptive liability scheme the facility where violating product is found, and all facilities upstream, are presumed liable for a violation of the diesel fuel standards. In this scenario, the truck loading terminal, the pipeline, and the facility of any other entity that distributed fuel in the terminal tank is liable. Each party, including the party owning the facility where the violation was discovered, has an opportunity to establish a defense to liability by meeting each of its defense elements. For most parties, to establish a defense the party must demonstrate that it did not cause the violation, that it had an appropriate quality assurance program (including periodic sampling and testing), and that product transfer documents account for the product and demonstrate that the fuel in question was apparently in compliance while in its custody. In addition, once the party discovers that the product is out of spec it must take appropriate actions to either redesignate the product to an appropriate designation or remedy the violation by blending in lower sulfur content fuel such that the resulting mixture meets the standard.

A terminal that discovers it has noncompliant product should immediately lock out the tank until it takes appropriate action. While the product is in violation until remedial action or downgrade, such action does allow the product to be transferred without additional violations. As in all of our other fuel programs, if you discover off-spec product and are taking some sort of remedial action, EPA will not issue a notice of violation (NOV). The product transfer documents and designation records must be changed to reflect the appropriate designation (if a change in designation is appropriate) and records should be kept to document the circumstances that gave rise to the violation and to document the remedial action or change in designation. If product is blended with other diesel fuel to bring it back within compliance with the 15 ppm sulfur standard, the product must be tested before it is transferred, to assure that such blending was successful.

Designate and Track Issues
For "designate and track" purposes, if a pipeline transfers product designated as 15 ppm to a terminal but it is subsequently discovered that the product did not meet the 15 ppm standard, then in addition to the standard violation there is also a designate and track violation. To remedy this violation, the parties must indicate in their records and on product transfer documents the appropriate revised designation at the time of transfer. In this example, the pipeline and

111

terminal must revise the designation of the fuel as it was transferred from the pipeline to the terminal. If the terminal elects to classify the product as 500 ppm highway fuel then, in this scenario, the pipeline must revise the designation for its hand-off to 500 ppm highway product and count the product against its anti-downgrade accounting.

The terminal may elect to upgrade the violating product (by blending it with product having a sulfur content less than 15 ppm) to meet the 15 ppm standard rather than downgrading the entire tank to 500 ppm fuel. If it does so, it may designate the resulting product as 15 ppm highway or nonroad fuel (but the designation of the product for the transfer from the pipeline to the terminal would be 500 ppm product).

If a terminal causes 15 ppm product to fail to meet the 15 ppm standard, and it conducts remedial action before shipment to bring the product to 15 ppm, for designate and track purposes, the product would still be counted as 15 ppm product when transferred to the terminal and 15 ppm product when transferred to the next facility. This is the case because designations are based on the product at the time that custody transfers from one facility to another, and not on the fuel's status while in storage. However, there would be a standard violation unless the terminal immediately changed classification of the product as soon as the contamination occurred (or immediately locked out the tank pending test results, then changed the classification or remedied the violation by blending in lower sulfur fuel).

13. GPA/Small Refiners

13.1:   EPA expanded the GPA (66 FR 19296). The GPA map on 66 FR 19300 is not consistent with the list of individual counties in Washington at 66 FR 19306. The map includes Klickitat County in the shaded area in Washington. However, the list of individual counties in Washington at § 80.215(a)(2)(i) does not include Klickitat County. Would you confirm that Klickitat County is in the GPA and your intention to include Klickitat County in a revised § 80.215(a)(2)(i)?

     A: In a direct final rule published on April 13, 2001 (66 FR 19296), we amended the Geographic Phase-in Area (GPA) to include counties and tribal lands in states adjacent to eight original GPA states. While Klickitat County appeared on the map on page 19300, it was inadvertently omitted from § 80.215(a)(2)(i) and should have been included in the list of GPA counties. This inconsistency was corrected in a recent action, which was published in the Federal Register on November 22, 2005 and can be found at: http://a257.g.akamaitech.net/7/257/2422/01jan20051800/edocket.access.gpo.gov/2005/pdf/05-22807.pdf).

13.2:   A refinery approved by EPA to produce gasoline subject to the interim GPA standards in 2007 and 2008 must demonstrate that by June 1, 2006 it will produce enough 15 ppm highway diesel fuel to meet the minimum 85 percent of its baseline volume requirement. See § 80.594(c). When is this demonstration due for GPA refineries? Is this demonstration part of another report (i.e., the 2003, 2004 and 2005 pre-compliance reports) or is it a separate, stand-alone submission?

     A: Under § 80.594(c), a refiner or importer approved to produce gasoline in the GPA subject to the gasoline sulfur standards under § 80.540 must demonstrate that by June 1, 2006 it will be producing a volume of on-road diesel fuel containing 15 ppm sulfur or less that is equal to or greater than 85% of its baseline volume (see § 80.540(e)). Under § 80.540(b) the demonstration of future production volume required under § 80.594(c) was due December 31, 2001, with the application under § 80.540 for approval of the gasoline sulfur extension. Under § 80.594(d), refiners and importers have until July 1, 2006 to submit a report stating that production or importation of 15 ppm sulfur motor vehicle diesel fuel will be started by June 1, 2006. The demonstration required by § 80.594(d) must be reported in a stand-alone report submitted by the refiner or importer.

13.3:   40 CFR 80.540(e) requires "The total volume of motor vehicle diesel fuel produced for use in the United States and designated as meeting the 15 ppm sulfur content standard under paragraph (d) of this section must meet or exceed 85% of the baseline volume established under paragraph (c) of this section, except that for the first compliance period from June 1, 2006 through June 30, 2007, the total volume must meet or exceed 92 percent of the baseline volume." Is it correct to assume that "produced" means available for sale at the refinery gate as an on-road fuel meeting 15 ppm diesel fuel standards?

A: *"Produced" should be interpreted as a refiner manufactured the motor vehicle diesel fuel from crude oil and the diesel fuel met all of the standards under § 80.520 and was designated as motor vehicle diesel fuel when it was delivered to the next facility in the distribution system.*

13.4: What records must be kept to demonstrate that an adequate volume has been produced?

A: *Under § 80.540(i), the record keeping requirements specified under § 80.592 must be met for GPA gasoline. This includes records indicating the volumes of each batch of diesel produced during the compliance period (§ 80.592(b)).*

13.5: How would a refinery with a loading rack account for 15 ppm diesel fuel that is sold into the offroad market? If production is interpreted in any other way, a refinery could be penalized if, due to logistical and production constraints, customers elect to purchase 15 ppm diesel fuel as an offroad diesel fuel.

A: *Per the requirements of § 80.598, promulgated in the nonroad diesel rule in June 2004, a refiner is responsible for designating the fuel appropriately when delivered to the next facility in the distribution system. Until June 2009, all 15 ppm diesel fuel must be designated as motor vehicle diesel fuel.*

13.6: Who are the small refiners in the ULSD program? Are they the same as the "small refiners" in the Tier 2 gasoline program?

A: *We have received 18 applications for small refiner status, and many of these refiners are currently approved small refiners for the gasoline sulfur and highway diesel programs. However, they may not all be approved for NRLM small refiner status. Specific information can be obtained upon request from Larry Haslett of EPA.*

13.7: If a large refinery spins off a small refinery how long does the small refiner have to apply for the small refiner status?

A: *The dates to apply for small refiner status were specified in both rulemakings, and the deadlines have passed for both highway and nonroad diesel.*

13.8: Is there an opportunity for a blender who is presently classified as a refiner to apply for small "refiner" status?

A: *The deadline to apply for small refiner status for both the highway and nonroad diesel fuel rules has passed. Furthermore, to be considered for small refiner status, a refiner must process diesel fuel from crude (see § 80.550 (a) and (b)).*

13.9: EPA indicated that the deadline for applying for small refiner status has already passed. However, should the marketer decide to blend e-diesel or biodiesel, it may want the option to be a "small refiner" under the diesel regulations.

A: *As stated in § 80.550 (a) and (b), the marketer would have to produce diesel fuel from crude.*

13.10: How do you establish a baseline for a small refiner, who has never produced on road diesel but wants to delay low sulfur gasoline equipment and build low sulfur diesel equipment first?

A: *It was discussed in the preamble to the final highway rule to allow case-by-case looking at a variety of factors (see pages 5073 and 5077, foot note). Unfortunately the provisions never made it into the regulations. We will be doing minor rule changes on this to mirror the preamble language.*

13.11: Is the decision to use Small Refiner Options 1, 2, or 3 irrevocable?

A: *Under § 80.551(c)(3), we ask for "an indication" of which kind of relief small refiners expect to use. The regulations do not prevent a refiner from using a different option later. However, section 80.553(h) voids the gasoline extension if small refiners fail to meet the 15 ppm volume requirements.*

13.12: Is an independent auditor required in order to establish a motor vehicle diesel baseline?

A: *Yes. As stated in § 80.595(c)(2), refiners must follow the procedures specified in §§80.91 through 80.93 (and use motor vehicle diesel fuel instead of gasoline) for the purposes of establishing a volume baseline under this section. Specifically, § 80.92 states that an independent auditor is required to establish a motor vehicle diesel baseline. Further, this baseline will be used to determine the baseline requirements set out in § 80.533.*

13.13: A company owns 58.75% of our refinery. The refinery is rated at 260,000 Barrels/day of which the company owns 152,750 B/D (260 * 0.5875). Would this qualify our refinery for the small refiner hardship provisions?

A: *That ownership percentage makes the refinery a subsidiary under our definition (see § 80.550) and all of a subsidiary's crude capacity must be used in determining the total crude capacity. There is no prorating.*

13.14: Can an approved highway diesel small refiner stay in business just by producing only 500 ppm diesel fuel and supply to the off road market.

> A: *Yes. An approved highway diesel small refiner that currently produces highway diesel can produce no highway diesel and use all of its diesel fuel production to supply the nonroad market.*

13.15: Our company is a small business refiner under the Small Business Administration (SBA) regulations. Our company would similarly meet the definition of a "small refiner" under the highway ULSD regulations, if not for the regulations' requirement that, for the purpose of determining qualification as a small refiner, a refiner must "include the employees and crude capacity of any subsidiary companies, any parent company and subsidiaries of the parent company in which the parent has 50 percent or greater ownership, and any joint venture partners." The distinction is attributable to the fact that, unlike the highway ULSD regulations, the SBA regulations do not affiliate our company with its owner.

Under the SBA regulations, our company's size is determined by counting the total capacity of all of our refineries, and the employees of all of our refineries and subsidiary distribution businesses. However, 13 CFR 121.103(b)(2) (SBA's regulations) exempts subsidiaries of our parent from affiliation, so that our company is not considered affiliated with the parent itself or other, unrelated subsidiaries of the parent. Our company believes that this definition is consistent with the purpose of the U.S. Environmental Protection Agency's affiliation rule, and is concerned that the provisions on affiliation in the highway ULSD regulations would inadvertently exclude our company.

> A: *Under EPA's gasoline and diesel sulfur regulations, we include all employees and crude capacity of all parent companies and subsidiaries where the parent company maintains 50 percent (or greater) ownership (§ 80.550(a) and (c)).*

13.16: My company has qualified for small refiner status and has chosen the small refiner options to produce 15 ppm highway and 15 ppm nonroad diesel fuel beginning June 1, 2006 (the "gas-for-diesel" options). However, the regulations state, at § 80.598(a)(3)(iv), that "prior to June 1, 2009 all 15 ppm sulfur MVNRLM diesel fuel must be designated as motor vehicle diesel fuel." Can EPA please clarify this?

> A: *This provision was revised with our May 2006 technical amendments to the regulations (see Section 16 for links to the technical amendments). Approved small refiners that choose the NRLM gas-for-diesel option only will be allowed to designate fuel as 15 ppm NRLM beginning June 1, 2006.*

13.17: The small refiner "gas-for-diesel" provision, § 80.553(b), states that "...starting no later than June 1, 2006, *all* motor vehicle diesel fuel produced by the refiner will comply with the 15 ppm sulfur content standard...." Shouldn't this say "95 percent"?

> A: *We intended to change this provision in the Nonroad Diesel rulemaking to state that 95% of the small refiner's diesel fuel must meet the 15 ppm standard (similar to the provision in § 80.540); this was amended by the May 1, 2006 technical amendments to the regulations (see Section 16 for links to all technical amendments).*

13.18: Small refiners and GPA refiners have the option to delay compliance with the Tier 2 gasoline requirements in exchange for compliance with the 15 ppm sulfur standard for highway diesel fuel beginning June 1, 2006 (the "gas-for-diesel" option). Under this option, starting June 1, 2006 at least 95% of the highway diesel fuel produced must meet the 15 ppm standard. How is compliance with the 95% requirement determined?

> A: *Compliance with the 95% requirements contained in § 80.553 (small refiners) and § 80.540 (GPA refiners) will be evaluated over each yearly compliance period (with the exception of the first compliance period for refinery production being a 13-month period). A refiner need not necessarily begin producing highway diesel fuel that meets the 15 ppm sulfur standard beginning June 1, 2006; however, 95% of the total volume of highway diesel fuel produced during any given compliance period must meet the 15 ppm sulfur standard. (For example- A refinery could choose to produce highway diesel fuel that meets 500 ppm sulfur for some short period of time initially; however, over the entire compliance period, a total 95% of the highway diesel fuel that the refinery produced must meet the 15 ppm sulfur standard. The highway diesel fuel that the refinery initially produced meeting the 500 ppm sulfur standard would be counted as part of the 5% highway diesel fuel that the refinery is allowed to produce during a compliance period that does not meet the 15 ppm sulfur standard.)*

13.19: The regulations at § 80.554(d) (the small refiner "gas-for-diesel" option) state that "95 percent of the NRLM diesel fuel produced by the refiner must be accurately designated...as meeting the 15 ppm sulfur standard..." ((d)(1)(i)) and "... the refiner must produce NRLM diesel fuel each year...that is equal to or greater than 85 percent of B_{NRLM}..." ((d)(1)(ii)). It was my understanding that for a small refiner that produces both highway and nonroad diesel fuel and chooses both the highway and nonroad diesel gas-for-diesel option, the small refiner would be required to produce 15 ppm fuel that is greater than or equal to a <u>combined</u> highway and nonroad baseline. Section 80.554(d), seems to state that small refiners would be required to meet separate highway and nonroad baselines, which could potentially force some small refiners to put more 15 ppm fuel into the NRLM market than they otherwise would have in order to produce at least 85 percent of their NRLM baseline.

A: *The regulations as they are currently written do state that at least 85 percent of the 15 ppm fuel that a small refiner produces must meet its NRLM baseline. However, upon further analysis of this provision, we believed that it would be more beneficial to require that small refiners meet or exceed 85 percent of a combined baseline ("$B_{MVNRLM} = B_{MV} + B_{NRLM}$"); this was amended by the May 1, 2006 technical amendments to the regulations (see Section 16 for a link to the technical amendments). Use of a combined baseline will allow small refiners to produce as much 15 ppm highway diesel fuel as possible rather than potentially being forced to put some 15 ppm diesel fuel into the NRLM market to be in compliance with the requirements of § 80.554(d).*

For small refiners that produce highway diesel fuel only, the combined baseline may also be used, as B_{NRLM} will be zero (and thus, B_{MVNRLM} will be equal to B_{MV}).

13.20: EPA has stated that it intended for the regulations to allow small refiners choosing the motor vehicle and/or NRLM "gas-for-diesel" options to produce 95 percent of the fuel that they produce to meet the 15 ppm sulfur standard. Can EPA please clarify how this fits in with the 85 and 92 percent figures that are already in the regulations?

A: *To use either, or both, of the gas-for-diesel provisions that are offered to small refiners, a small refiner must apply for a baseline. The 85 percent requirement simply requires that the minimum absolute volume of 15 ppm diesel fuel that a small refiner produces must be 85 percent of its baseline (however, for the compliance period from 6/1/06 through 6/30/07 a small refiner must meet or exceed 92 percent of its baseline). The 95 percent requirement requires that, of a small refiner's total annual production, at least 95 percent of that fuel must meet the 15 ppm sulfur standard.*

For example-
A small refiner elects to use the highway diesel gas-for-diesel option. Per, the requirements of § 80.553(b), the refiner must apply for a motor vehicle diesel fuel volume baseline. The small refiner produced 300 gallons of fuel in both 1998 and 1999, so its motor vehicle diesel fuel baseline is 300 gallons (per § 80.596, the motor vehicle diesel fuel volume baseline is calculated as the annual average of the refiner's production volumes from January 1, 1998 through December 31, 1999). Thus, the refiner's yearly production of 15 ppm motor vehicle diesel fuel must be at least 85 percent of its baseline, or 255 gallons. In addition, 95 percent of the total annual volume of motor vehicle diesel fuel that the small refiner produces must meet the 15 ppm sulfur standard, and this volume of 15 ppm fuel must meet or exceed 255 gallons. So, if the small refiner's total annual production of motor vehicle diesel fuel is 150 gallons, it could not qualify for the gas-for-diesel option, because its required volume of 15 ppm production (143 gallons) would not meet or exceed 85 percent of its baseline. However, the small refiner could qualify if its total annual production of motor vehicle diesel fuel was

270 gallons, since the required amount of 15 ppm motor vehicle diesel fuel (257 gallons) exceeds 85 percent of its baseline.

13.21: All sections of the regulations that apply to GPA and small refiners (such as § 80.540 and § 80.552) it states that the annual compliance period is June 1, 2006 through June 30, 2007. However, § 80.599(a)(1) states that the annual compliance period is June 1, 2006 through May 31, 2007. Which are the correct dates for the first compliance period?

    A: *As stated above in QA 4.44, the first annual compliance period for refinery production (for compliance with the Temporary Compliance Option, small refiner, and GPA refiner provisions) is June 1, 2006 through June 30, 2007. Therefore, all sections regarding compliance for small and GPA refiners state June 30, 2007 as the end of the first compliance period. Section 80.599 lists the D&T quarterly and annual compliance periods. The first annual compliance period for D&T ends May 31, 2007 (which corresponds with the start date of the NRLM program on June 1, 2007).*

          All parties in the D&T system are required to report under D&T, and therefore shall determine their compliance with the annual compliance periods listed in § 80.599.

          Refiners and importers must also show compliance with the 80/20 production requirements (and GPA and small refiners must show this compliance to remain eligible for the GPA and small refiner flexibility provisions) and thus shall determine compliance with these using June 1, 2006 through June 30, 2007 (as stated in § 80.530) as the first annual compliance period.

119

14.   Transmix/Interface Fuel

14.1:   In the case where a terminal processes transmix with an on-site fractionator, is the on-road production volume subject to the requirements of § 80.520?

> *A: The regulations finalized with the nonroad rule modified the requirements for transmix processors who do not also produce fuel from crude. These provisions, contained in § 80.513, allow for 100 percent of a transmix processor's highway production to remain at the 500 ppm sulfur level until June 1, 2010.*

14.2:   Interfaces between jet fuel and other distillates are cut into the other distillate to protect the jet fuel, thereby increasing the volume of the other distillate. Kerosene/15 ppm diesel fuel and high sulfur diesel/15 ppm diesel fuel interfaces could be cut into separate tankage to create 500 ppm on-road diesel (assuming no dye in the high sulfur diesel). Is there any limit as to the amount of 500 ppm on-road diesel that could be created through these interface cuts? Are there any special downstream requirements for the testing or documentation of on-road diesel created from interface?

> *A: Under § 80.527(c), the maximum allowable amount of #2 15 ppm on-road diesel that can be downgraded to 500 ppm on-road diesel is 20%, otherwise there are no downgrade restrictions. However, on-road diesel created from interface must meet the standards under § 80.520.*

14.3:   How will EPA treat transmix, which does not fit into any category? It is not on-road and it is not off-road gas or fuel. Transmix is not intended for final use by any consumer but it can be re-refined or re-processed.
    a)   For EPA purposes, will it need to be tracked after its creation?
    b)   If it is sold to a refinery for re-refining, will EPA treat it differently than if it is sold to a reprocessor?
    c)   Does it make a difference if it is sold to another refiner or the original refiner?

> *A:*
> *a)   Transmix would merely be designated as such and would not be tracked for EPA purposes unless (and until) it is redesignated for fuel use.*
> *b)   It would not be treated any differently. However, transmix processors who are also refiners are held to different standards for the fuel that they produce from transmix.*
> *c)   No, there is no difference if it is sold to another, or the original, refiner.*

14.4:   Can a terminal tank meet the 15 ppm diesel fuel sulfur standard on average, or must the diesel fuel meet the 15 ppm sulfur standard on a per-gallon basis? For example, if a pipeline or terminal accidentally cut some of an interface, consisting of 500 ppm diesel fuel and 15 ppm diesel fuel, into the 15 ppm diesel fuel, causing a certain strata of a 15 ppm diesel fuel storage tank to exceed the 15 ppm standard, the rest of the tank may still contain product that continues to meet the 15 ppm standard. In the alternative, it may be possible to remedy a violation by blending diesel fuel exceeding 15 ppm sulfur content by blending sufficient quantity of very low sulfur product to make the tank or batch meet the 15 ppm standard. Would EPA allow such remedial action?

A: *Diesel fuel subject to the 15 ppm sulfur standard must meet that standard on a per-gallon basis and not on average. The pipeline or terminal must handle the interface in a manner that ensures that product distributed as subject to the 15 ppm cap is in compliance with that cap. If any portion of a tank that is offered for sale or supply as fuel subject to the 15 ppm sulfur cap is out of compliance with the 15 ppm cap, then that volume of fuel would be in violation. However, if a pipeline or terminal accidentally cuts part of an interface having a sulfur content greater than 15 ppm into a tank containing fuel subject to the 15 ppm sulfur standard, it may be possible to remedy the violation. We would not pursue an enforcement action where a terminal determines that product exceeding the 15 ppm standard has been added to a tank if the terminal: discovers the commingling of higher sulfur product with 15 ppm product through its quality assurance program and not by an EPA inspection; immediately locks down the tank before any of the product is released; transfers product exceeding the 15 ppm standard to some appropriate high sulfur product distribution; retests the 15 ppm product tank to assure that the entire remaining volume of the tank meets the 15 ppm standard before releasing the product as 15 ppm product; and maintains records demonstrating the occurrence, the actions taken to remedy the violation, sampling and test results, and actions to prevent future violations.*

Such situations may also be remedied by blending the contaminated product with low sulfur product in order to bring the sulfur level of all the product in the tank to the 15 ppm sulfur standard. Again, before releasing product the sulfur level should be retested, and records of the occurrence, remedial actions, sampling and test results, and of actions to prevent future violations should be maintained.

Finally, such situations could be remedied by downgrading the entire volume of product affected by the commingling with the higher sulfur interface. When 15 ppm product is downgraded, the party should document the circumstances that gave rise to the downgrading. The diesel fuel in question should be segregated from diesel fuel subject to the 15 ppm standard, the product transfer documents must reflect the downgraded classification, and the diesel fuel must not be sold, dispensed or transported in a manner that is inconsistent with the downgraded classification.

14.5: The facilities that my company owns include a pipeline, terminal, and refinery. We wish to blend the transmix generated due to normal pipeline operations into a barge containing crude oil at our terminal facility. This crude oil and transmix mixture would then be taken to our refinery for processing. Is there anything in the clean diesel regulations that places restrictions on this practice?

> A: There is nothing in the highway or nonroad diesel fuel regulations to prevent this practice from continuing. The losses to transmix would merely be reflected in the pipeline/terminal aggregated facility's report.

14.6: Could the < 500 ppm sulfur diesel fuel that a transmix processor manufactures from reprocessing transmix obtained from a refinery be sold into the highway diesel market until 2010? Could a transmix processor sell <500 ppm sulfur diesel fuel it manufactures from tank bottoms into the 500 ppm highway diesel market?

> A: 40 CFR 80.513, the provisions that apply to transmix processing facilities states that the term transmix applies to a mixture of finished fuels that no longer meets the specifications for a fuel that can be used or sold without further processing. For the mixture to be considered transmix under section § 80.513, the component finished fuels that make up this mixture must have been previously certified as compliant with the applicable compositional requirements and the mixing must have been non-intentional. The volume of <500 ppm diesel that a transmix processor manufactures from reprocessing this transmix could be sold into the 500 ppm highway diesel market until May 31, 2010. Any volume of < 500 ppm sulfur diesel fuel that a transmix processor manufactures from feedstock (e.g. a diesel blendstock such as light cycle oil) that does not meet the definition of transmix contained in section § 80.513 would have to meet the 80/20 requirements. Tank bottoms are the remains of previously certified finished fuels and as such would meet the definition of transmix under section § 80.513. Thus, <500 ppm diesel fuel manufactured using tank bottoms by a transmix processor could be sold into the 500 ppm highway diesel market until May 31, 2010.

14.7: My company operates a pipeline. We currently blend small quantities of transmix into diesel fuel subject to a 500 ppm sulfur standard during shipment. The quantities blended have a negligible impact on the sulfur content of the fuel. Is there anything in the new highway and nonroad diesel sulfur programs that would prevent us from continuing to blend transmix into diesel fuel subject to a 500 ppm standard in the manner we do today? The blending of transmix into the 500 ppm diesel would result in an increase in the total highway diesel volume. How would this be accounted for in determining compliance with the highway diesel volume balance requirements?

> A: EPA has not previously considered the blending of transmix into diesel fuel subject to a 500 ppm sulfur standard during shipment by pipeline. The diesel sulfur regulations do not specifically address blending transmix into diesel fuel. We believe that it would be appropriate to provide this flexibility, provided the pipeline operator met the following conditions:

- *The transmix blended into 500 ppm diesel fuel would need to result from factors associated with normal pipeline operations.*
- *The pipeline would need to conduct periodic tests of the 500 ppm diesel fuel into which transmix was blended which show that the transmix blending would not cause the 500 ppm sulfur cap to be exceeded, and that the diesel fuel still met all the applicable ASTM standards for diesel fuel such as flash point or distillation points.*
- *The pipeline would need to retain documents that reflect the rate of transmix blending and make these documents available to EPA upon request.*

The volume of transmix blended into 500 ppm highway diesel fuel would need to be accounted for in demonstrating compliance with the highway diesel volume balance calculations. The pipeline would have to account for the volume of transmix blended into highway diesel fuel by including the transmix volume as 500 ppm diesel production in the compliance calculations in sections § 80.599, § 80.600 (except for the refiner batch reports, new volumes should simply be accounted for in the D&T calculations) and § 80.601 of the diesel sulfur regulations.

14.8: How should pipelines characterize "interface fuel" (slightly off-spec fuel that may be created if midpoint batch cuts are made) on PTDs when delivering fuel to a terminal?

A: *Q&A 2.47, above, discusses the types of language that EPA would allow on a PTD to account for such situations.*

15. General/Miscellaneous

15.1: Why is heating oil exempt from the sulfur content regulations? If it's bad for cars, why not for people's homes?

> *A:* *The Clean Air Act does not give EPA specific authority to regulate heating oil.*

15.2: Will there be any significant quantities of kerosene that get hydrotreated to 15 ppm (or lower) sulfur for making ULSD?

> *A:* *We did not require early reporting of this information in the pre-compliance reports, and as such do not have specific data to address it. Nevertheless, we are currently aware that at least two refiners plan on manufacturing 15 ppm #1 diesel fuel for wintertime blending purposes.*

15.3: What does EPA intend to do with all of these reporting requirements and paperwork to justify their cost to the consumer and taxpayer?

> *A:* *The D&T and recordkeeping provisions were put in place as a means of reducing the capital and operating costs otherwise necessary to segregate fuels. The costs of these provisions were estimated in the draft Information Collection Request (ICR) submitted to OMB, and are small in comparison to the overall costs and benefits of the program.*

15.4: *[This Q&A has been revised and is now question 5.3.]*

15.5: Does "other sources of volume" refer to diesel desulfurizers? What do you mean by "produced from crude oil"?

> *A:* *"Other diesel fuel" means any distillate products that meet the definition of diesel fuel, such as kerosene, that is represented to have a sulfur content less than or equal to 500 ppm. "Produced from crude oil" means that the refiner manufactured the diesel fuel from crude oil, typically through distillation and hydroprocessing.*

15.6: If terminals opt to wash barges, tanks, equipment and trucks before running 15 ppm diesel fuel through their systems, the flushing process will generate substantial volumes of oily-water waste. Has the EPA given thought to assisting the industry in dealing with this waste?

> *A:* *The highway diesel rulemaking did not specifically address this issue. The rulemaking allowed terminals until July 15, 2006 (approximately 6 weeks after the June 1, 2006 compliance date for refiners and importers), before any 15 ppm*

highway diesel fuel they distribute must meet the 15 ppm standard. A technical amendment to the rulemaking (published November 22, 2005) will now allow an additional 45 days– until September 1, 2006– before highway diesel fuel that terminals distribute must meet the 15 ppm sulfur standard. This additional time should help to minimize waste.

15.7: Public reaction to new fuels in the past has been mixed (i.e., RFG in Milwaukee, CARB II Diesel in California). What steps does EPA intend to take to educate and inform owners of diesel-powered motor vehicles of the changes to fuel specifications, performance issues, and/or air quality benefits?

> *A: EPA currently has all documents associated with the diesel fuel rules available on its website (http://www.epa.gov/cleandiesel/) and will be periodically providing additional updates to the website to provide summary descriptions for the public of the benefits and impacts of the rules. Additional steps will be determined over the course of the remaining months leading up to June 2006. EPA is participating in an outreach working group with industry and other governmental agencies on developing outreach materials to educate the public on ULSD- the Clean Diesel Fuel Alliance (www.clean-diesel.org).*

15.8: In the past, EPA has, with the assistance of industry trade associations, produced fuels brochures for use at retail outlets explaining new fuel programs and addressing performance and air quality issues. Is EPA planning a similar effort with respect to the diesel fuel programs?

> *A: In November 2005, the Clean Diesel Fuel Alliance was formed. Many public and private organizations are collaborating through the Clean Diesel Fuel Alliance to facilitate the introduction of ULSD. The U.S. Department of Energy (DOE), EPA, engine, vehicle and component manufacturers, all sectors of the petroleum industry, and fuel consumers, such as truckers, are providing comprehensive information and technical coordination. The CDFA has developed brochures, handouts, and an FAQ on ULSD. Please visit the CDFA website for more information at www.clean-diesel.org.*

15.9: How will the highway diesel rule affect mobile refuelers ("wet hosers")?

> *A: Mobile refuelers will be subject to the prohibitions against misfueling like a retailer. However, we do not require them to label their truck storage tanks since the product may change. They would also be liable for contaminating any 15 ppm fuel such that it no longer met the standard. They are required to deliver an accurate product transfer document to the customer.*

15.10: Fuel pump labels for LSD (500 ppm highway diesel fuel) state that "Federal law prohibits use in model year 2007 and later highway vehicles and engines." I have a model year 2007 vehicle that is equipped with a model year 2006 engine, can I use 500 ppm highway diesel fuel or is 15 ppm highway diesel required to be used in my engine?

A: *Many 2007 model year trucks (those produced prior to Jan 2007) will in fact have model year 2006 engines which may lawfully be fueled with either 500 ppm highway diesel fuel or 15 ppm highway diesel fuel. We interpret the reference to "model year 2007 vehicles" in our fuel pump labeling and misfueling regulations as referring to those model year 2007 vehicles where either the vehicle or engine is certified to the model year 2007 emissions standards. Vehicles with engines certified to 2006 model year standards would generally not be using these sulfur sensitive emission control systems. Through amendments made to the vehicle labeling regulations, the fueling of model year 2007 vehicles equipped with model year 2006-certified diesel engines (excluding those engines that certified early to the 2007 standards) with 500 ppm sulfur diesel fuel is permitted, as long as EPA has approved a certificate of conformity applicable to the engine and not the vehicle. Further, any vehicle with a model year 2007 engine/with an engine that is certified to 2007 model year standards will have a label which states that ULSD must be used. Please see the Clean Diesel Fuel Alliance website (www.clean-diesel.org) to view a letter from EPA to the American Trucking Associations and the Engine Manufacturers Association that addresses this issue.*

16.   Key Regulatory Documents

☞   **Link to the Electronic Code of Federal Regulations (eCFR):**
*The "eCFR" is a compiled version of all of the regulations that is updated weekly.
Please note that if a regulation has a pending amendment that has not yet become
effective, it will be stated at the top of the page (e.g., "Link to an amendment published at
12 FR 34567") along with a direct link to that amendment.*
http://ecfr.gpoaccess.gov/cgi/t/text/text-idx?c=ecfr&sid=94cf84cc9b9f6927a1ff4daf8d7f8
642&tpl=/ecfrbrowse/Title40/40cfr80_main_02.tpl

☞   Highway Diesel Rulemaking (*published January 18, 2001*):
Preamble-
http://www.epa.gov/otaq/regs/hd2007/frm/frdslpre.pdf
Regulations-
http://www.epa.gov/otaq/regs/hd2007/frm/frdslreg.pdf

☞   Nonroad Diesel Rulemaking (*published June 29, 2004*):
Preamble & Regulations-
http://www.epa.gov/otaq/url-fr/fr29jn04.pdf

☞   July 2005 Technical Amendment to the Highway and Nonroad Diesel Rules (*published
July 15, 2005*):
Preamble & Regulations-
http://a257.g.akamaitech.net/7/257/2422/01jan20051800/edocket.access.gpo.gov/2005/pdf/05-13781.pdf

☞   November 2005 Technical Amendment to the Highway and Nonroad Diesel Rules
(*published November 22, 2005*):
Preamble & Regulations-
http://a257.g.akamaitech.net/7/257/2422/01jan20051800/edocket.access.gpo.gov/2005/pdf/05-22807.pdf

☞   April/May 2006 Technical Amendment to the Highway and Nonroad Diesel Rules
(*published May 1, 2006*):
Preamble & Regulations-
http://a257.g.akamaitech.net/7/257/2422/01jan20061800/edocket.access.gpo.gov/2006/pdf/06-3930.pdf